Analog Circuits and Signal Processing

Series Editors:

Mohammed Ismail, Dublin, USA

Mohamad Sawan, Montreal, Canada

The *Analog Circuits and Signal Processing book series*, formerly known as the *Kluwer International Series in Engineering and Computer Science*, is a high level academic and professional series publishing research on the design and applications of analog integrated circuits and signal processing circuits and systems. Typically per year we publish between 5–15 research monographs, professional books, handbooks, edited volumes and textbooks with worldwide distribution to engineers, researchers, educators, and libraries.

The book series promotes and expedites the dissemination of new research results and tutorial views in the analog field. There is an exciting and large volume of research activity in the field worldwide. Researchers are striving to bridge the gap between classical analog work and recent advances in very large scale integration (VLSI) technologies with improved analog capabilities. Analog VLSI has been recognized as a major technology for future information processing. Analog work is showing signs of dramatic changes with emphasis on interdisciplinary research efforts combining device/circuit/technology issues. Consequently, new design concepts, strategies and design tools are being unveiled.

Topics of interest include:
Analog Interface Circuits and Systems;
Data converters;
Active-RC, switched-capacitor and continuous-time integrated filters;
Mixed analog/digital VLSI;
Simulation and modeling, mixed-mode simulation;
Analog nonlinear and computational circuits and signal processing;
Analog Artificial Neural Networks/Artificial Intelligence;
Current-mode Signal Processing;
Computer-Aided Design (CAD) tools;
Analog Design in emerging technologies (Scalable CMOS, BiCMOS, GaAs, heterojunction and floating gate technologies, etc.);
Analog Design for Test;
Integrated sensors and actuators;
Analog Design Automation/Knowledge-based Systems;
Analog VLSI cell libraries;
Analog product development;
RF Front ends, Wireless communications and Microwave Circuits;
Analog behavioral modeling, Analog HDL.

More information about this series at http://www.springer.com/series/7381

Francesco Mazzilli • Catherine Dehollain

Ultrasound Energy and Data Transfer for Medical Implants

Francesco Mazzilli
Melexis Technologies SA
Bevaix, Switzerland

Catherine Dehollain
ELB 231 Batiment ELB
EPFL SCI STI CD
Lausanne, Vaud, Switzerland

ISSN 1872-082X　　　　　　　　ISSN 2197-1854　(electronic)
Analog Circuits and Signal Processing
ISBN 978-3-030-49006-5　　　　　ISBN 978-3-030-49004-1　(eBook)
https://doi.org/10.1007/978-3-030-49004-1

This Springer imprint is published by the registered company Springer Nature Switzerland AG
The registered company address is: Gewerbestrasse 11, 6330 Cham, Switzerland

If we knew what it was we werse doing, it would not be called research, would it?

—*Albert Einstein*

I dedicate this book to my parents, sister and my girlfriend Chloé with love and gratitude.

Preface

The technology evolution in the last decade has promoted the development of new sensing and monitoring devices for healthcare. The upcoming great challenge for diagnosis devices lies in their ability to continuously monitor a patient's physical and biochemical parameters, under the natural physiological status of the patient, and in any environment. Human body monitoring using a network of wireless sensors may be achieved either by attaching these sensors to the body surface (wearable medical devices) or by implanting them into tissues (implantable medical devices), depending on their functionality.

This book proposes systems and methods, not using RF or magnetic telemetry with their crippling limitations, but acoustic waves (ultrasonic), for communicating with and energizing efficiently a network of transponders that are placed deep within a human body. Therefore, the primary key objective is to develop a novel telemetry technology that will enable any kind of device implanted deep inside the body to communicate wirelessly with an external system, providing patients and physicians with valuable tools, respectively, to continuously monitor and treat several potentially life-threatening conditions in a minimally invasive manner.

The implantable transponder contains an energy exchanger which converts acoustic energy into electrical energy, a small local energy storage (rechargeable microbattery), a control and processing chip (low-power microcontroller), and a simple transceiver. This book also incorporates a dedicated section on the external control unit, capable of powering and receiving information directly from the implanted device via ultrasound.

A discrete version of the implanted device is realized to test energy power transmission and characterize the uplink communication while changing the distance between the control unit and the implant. The microcontroller ATMEGA 48A handles the different operations as the battery reaches enough charge. For the uplink, the backscattering communication principle is applied to ultrasound waves to remove the use of a local oscillator in the implanted device, thus minimizing power consumption.

A second version of the implanted device is realized using a standard CMOS process. Hence, a novel active rectifier is designed in 0.18 μm HV CMOS process which presents 80% power conversion efficiency and a chip area of 0.114 mm^2. To suppress voltage noise on the supply rail, a two-stage low-drop-out regulator is designed with the same CMOS process aimed to recharge a Lithium Phosphorus Oxynitride (LiPON) microbattery. The downlink communication is studied and a non-coherent demodulation scheme is adopted, thus a variable gain amplifier is designed in 0.18 μm low-power CMOS process, which shows a maximum of 58.8 dB differential gain at 1 MHz and a bandwidth of 2.5 MHz.

The research leading to these results has received funding from the European Community's Seventh Framework Programme (FP7/2007-2013) under grant agreement n. 224009.

Bevaix, Switzerland   Francesco Mazzilli

Lausanne, Switzerland   Catherine Dehollain

Acknowledgements

I thank Dr. Catherine Dehollain for the support, confidence, and academic freedom she has provided to carry out this work.

I thank all my friends and former colleagues who contributed to make pleasant the several years I spent at EPFL while I was working on ultrasound. For many years after my graduation as PhD student from EPFL, we stayed in touch. I greet my valuable friends: Charly, our discussions were endless looking for the meaning of life, who is now living in Israel with his wife and son. Enver and Oguz, for introducing me to the Turkish culture; they left Switzerland to pursue their career in Canada and in the USA. Arnab, an explosive source of energy, who is working in Germany and Luca, the break of four o'clock, who became a manager. An enormous thanks goes to my flatmates, Massimo and Andrea, our weekly homemade pizza and parties will be unforgettable. A special thanks goes to Vincent Praplan, a brilliant engineer who helped me in the design of the hardware for the external base station.

I wish to thank my parents, my sister, my brother-in-law, and my girlfriend Chloé for their support and love.

Lausanne, Switzerland Francesco Mazzilli
February 2020

Contents

About the Authors

Francesco Mazzilli is an analog design engineer at Melexis Technologies SA in Chemin Bevaix, Switzerland.

Catherine Dehollain received the Master Degree in Electrical Engineering in 1982 from EPFL. Then, she worked in Geneva up to 1990 as a Senior Design Engineer in telecommunications at the European research center of Motorola. From 1990 up to 1995, she did her PhD thesis at the Chaire des Circuits et Systemes at EPFL in the domain of impedance broadband matching circuits. Since 1995, she is responsible at EPFL for the RFIC group. She has participated in different Swiss research projects as well as European projects dedicated to data communication of sensors nodes (e.g. MuMoR, Minami European projects) as well as remote powering of sensor nodes. Her main domains of interest are telecom applications (e.g. impulse radio ultra-wide band, super-regenerative receivers, RFIDs) as well as biomedical applications. She has been the coordinator of European projects (e.g. FP6 SUPREGE, FP7 Ultrasponder)and of Swiss projects (e.g. CAPED CTI project, NEURO-IC SNF project).

List of Acronyms

AM	Amplitude modulation
ASK	Amplitude-shift keying
BER	Bit error rate
BW	Bandwidth
CDRH	Center for Devices and Radiological Health
CMOS	Complementary metal-oxide semiconductor
CU	Control unit
CW	Continuous wave or continuous waveform
EH	Energy harvesting
FDA	Food and Drug Administration
FPGA	Field-programmable gate array
GBW	Gain-bandwidth (product)
HV	High voltage
IC	Integrated circuit
IMD	Implantable medical device
I_{SPPA}	Spatial-peak pulse average intensity
I_{SPTA}	Spatial-peak temporal average intensity
LDO	Low-drop-out
LiPON	Lithium phosphorus oxynitride
LNA	Low-noise amplifier
LSK	Load shift keying
Mbps	Megabit per second
MI	Mechanical index
MOSFET	Metal-oxide semiconductor field-effect transistor
OOK	On–off keying
PA	Power amplifier
PD	Pulse duration
PLL	Phase-locked-loop
PMU	Power management unit
PSRR	Power supply rejection ratio
RF	Radio frequency

RFID	Radio frequency identification
RX	Receiver
TR	Transponder
TX	Transmitter
UHF	Ultra high frequency
US	Ultrasound
VGA	Variable gain amplifier
WBAN	Wireless body area network
WPT	Wireless power transfer
WuRX	Wake-up receiver

List of Figures

List of Tables

Chapter 1
Introduction

Keywords Ultrasound · Implanted medical device (IMD) · Wireless power transfer (WPT) · Energy harvesting · Health monitoring · Inductive coupling · Transducer · Energy storage · Battery · Radio frequency identification (RFID) · Control unit (CU) · Analog-to-digital converter (ADC) · Digital signal processor (DSP) · Attenuation · Frequency

In the 1950s ultrasound started to be used for biomedical imaging and for therapy purposes [1] and a study published in 2001 has shown the ability of acoustic waves to energize and transmit information within the human body [2]. To this end new implanted medical devices (IMDs) are appearing among health monitoring applications [3] which exploit the capability of acoustic waves to penetrate deeper in the body tissue without being significantly attenuated [4, 5], and inherently to avoid interference with other medical systems.

Inductive coupling is a valid alternative to ultrasound for near field energy transmission. A theoretical comparison between the ultrasound and the inductive power delivery method has been presented [6] showing the wireless power transmission (WPT) efficiency at different distances. Inductive coupling performs better for shorter distances between the transmitter and receiver, while ultrasonic coupling performs better for larger distances. The distance at which the acoustic link outperforms the inductive link depends on the receiver size. For a receiver with a diameter equal to 10 mm, the acoustic link performs better when the distance is above 3 cm. If the diameter of the receiver is set to 5 mm, then the distance cross point is 1.5 cm. The optimum power transfer depends on the operating frequency, for the acoustic system a carrier frequency in the range of 1 MHz was used, while for the inductive link was fixed to 13.56 MHz.

The requirements for IMDs deeply (e.g. deeper than 5 cm) introduced into the human body are ruled by the implant size (1–3 cm^3) and by its longevity, which is determined by the energy storage capabilities [7], and by the modulator power consumption within the transponder. Therefore, a load modulation technique (backscattering modulation), largely exploited in radio frequency identification (RFID), may be the best candidate for medical devices due to its low-power

© Springer Nature Switzerland AG 2020
F. Mazzilli, C. Dehollain, *Ultrasound Energy and Data Transfer for Medical Implants*,
Analog Circuits and Signal Processing, https://doi.org/10.1007/978-3-030-49004-1_1

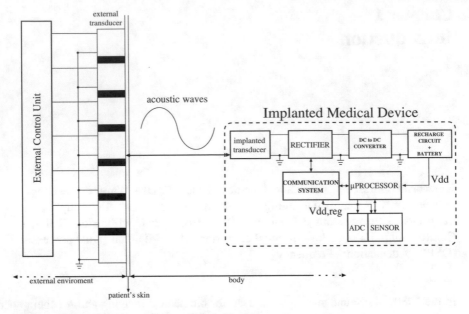

Fig. 1.1 Schematic of a complete transponder device

consumption [8]. However, IMDs should be tested in-vitro for primary analysis of side effects so that this new technology can be accepted for clinical use.

Hence, a continuous monitoring system such as an IMD with processing and wireless communication could help a patient suffering from chronic diseases. Heart rhythm abnormalities such as atrial fibrillation are commonly encountered in clinical practice and regular monitoring is required to ensure control of the heart rate. Hypertension is another cardiovascular disease and an intelligent IMD, such as an ultrasonic transponder system, would enable physicians to monitor patients with high blood pressure during normal activity. Also elderly patients, a group whose life expectancy is continuously increasing, can benefit of the ultrasonic transponder system. In this case identifying ways of surveying in a minimally invasive manner in their domestic environment is very important especially during months of non-temperate weather (either very cold or very hot). This would allow earlier detection of any degradation in their condition.

Figure 1.1 shows the main building blocks of the ultrasound link. An external control unit drives a transducer array to transmit data and energy towards an implanted medical device. Hence, a second transducer is present to convert the acoustic into electric energy by an electroacoustic transducer either to recharge a microbattery or to upload/download information. In order to monitor an activity in the human's body a customized sensor is needed along with an analog-to-digital converter (ADC) and a digital signal processor (DSP) in order to process and to collect the data.

1.1 Book Outline

In Chap. 2, ultrasound technologies in medicine are described. An example is shown on how ultrasound can be employed for remote powering and wireless communication, thus an overview to the project is given.

In Chap. 3, exposure limits to ultrasound for human tissues are introduced so that system specifications can be derived.

In Chap. 4, the architecture of the control unit is described. The first part deals with designed of the power amplifier aimed for the transmission of power and data towards the implant. In the second part, the receiver to retrieve the data from the implant is presented.

In Chap. 5, the architecture of the transponder is addressed. First, a narrow band equivalent model of the piezoelectric transducer is introduced and validated through measured data. This model is used to design the integrated AC-to-DC, and a comparison on various rectifiers architectures proposed in the literature is presented. The design of a two-stage linear regulator is demonstrated to reduce voltage noise on the supply rail which is present at the output of the rectifier. An amplitude-shift keying demodulation scheme is proposed, and the design of the front-end amplifier is discussed in details. The chapter ends presenting the modulation scheme, based on load modulation.

In Chap. 6, the proposed architectures are combined in two prototypes to demonstrate separately wireless power transfer along with data transfer. A discrete version of the implant is used to demonstrate the modulation scheme, while an integrated implementation in a 0.18 μm technology demonstrates the demodulation scheme plus the integrated battery charger.

Lastly, the conclusion and outlook is presented in Chap. 7.

References

1. J. Aldes, W. Jadeson, S. Grabinski, A new approach to the treatment of subdeltoid bursitis. Am. J. Phys. Med. **33**, 79–88 (1954)
2. H. Kawanabe, T. Katane, H. Saotome, O. Saito, K. Kobayashi, Power and information transmission to implanted medical device using ultrasonic. Jpn. J. Appl. Phys. **40**(Part 1, No. 5B), 3865–3866 (2001)
3. B.C. Tran, B. Mi, R.S. Harguth, Systems and methods for controlling wireless signal transfers between ultrasound-enabled medical devices, 2009
4. T. Yamada, T. Uezono, K. Okada, K. Masu, A. Oki, Y. Horiike, RF attenuation characteristics for *in vivo* wireless healthcare chip. Jpn. J. Appl. Phys. **44**(7A), 5275–5277 (2005)
5. A. Ifantis, A. Kalis, On the use of ultrasonic communications in biosensor networks, in *BIBE*, 2008, pp. 1–6
6. A. Denisov, E. Yeatman, Ultrasonic vs. inductive power delivery for miniature biomedical implants, in *2010 International Conference on Body Sensor Networks (BSN)*, June 2010, pp. 84–89
7. X. Wei, J. Liu, Power sources and electrical recharging strategies for implantable medical devices. Front. Energy Power Eng. Chin. **2**, 1–13 (2008)
8. J.-P. Curty, N. Joehl, C. Dehollain, M.J. Declercq, Remotely powered addressable UHF RFID integrated system. IEEE J. Solid-State Circuits **40**(11), 2193–2202 (2005)

Chapter 2
Ultrasound in Medicine

Keywords Ultrasound · Pulse-echo · Rectifier · Implanted medical device (IMD) · Wireless power transfer (WPT) · Wireless energy transfer · Food drug administration (FDA) · Wireless communication · Energy harvesting · Health monitoring · Inductive coupling · Transducer · Battery · Control unit (CU) · Attenuation · Frequency

The technology evolution in the last decade has promoted the development of new sensing and monitoring devices for healthcare. However, the design issues for deeply implanted devices remain numerous, especially miniaturization and power consumption challenges, which are related. In addition, because of the body's dielectric nature, communicating with implants that are located deep within the body, using conventional techniques like Radio Frequency (RF), may also not work effectively. To overcome these limitations, exclusive technologies based on ultrasonic telemetry techniques are developed. In this chapter, a global overview on ultrasound technology and its use in medical environment is given. Moreover, an example is shown on how ultrasound can be employed for remote powering and wireless communication.

2.1 Ultrasonic Techniques

The non-invasive property of ultrasound and its ability to distinguish interfaces between tissues of different acoustical impedance has been its main attraction as a diagnostic procedure [1]. One of the most useful applications of ultrasound has been echo-encephalography, which allows position measurement of prominent brain structures. Moreover thanks to obstetrics, ultrasound can be used to determine the presence of one or more fetuses, while in gynecology, intrauterine tumors and cysts may be detected. Echocardiography is another ultrasound diagnostic methodology that provides information on the size and the shape of the heart, pumping capacity, and the location and extent of any tissue damage. Hence, ultrasonic diagnostic

© Springer Nature Switzerland AG 2020
F. Mazzilli, C. Dehollain, *Ultrasound Energy and Data Transfer for Medical Implants*,
Analog Circuits and Signal Processing, https://doi.org/10.1007/978-3-030-49004-1_2

techniques are mainly classified in three categories: (1) pulse-echo, (2) Doppler, and (3) acoustic imaging. Since pulse-echo technologies are applicable to both Doppler and holographic domains, most of the discussion will be on developments in this area.

2.1.1 Pulse-Echo Ultrasound

Pulse-echo developments are divided into three categories: (1) display techniques, (2) resolution improvement techniques, and (3) processing techniques. Improvements in display techniques are related to the sensitivity of the equipment (e.g. oscilloscope) that has to show reflections measured via the acoustic transducer. Then, the image can be represented either in gray scale or in colored presentation. Resolution improvement techniques are largely dependent on the characteristics of the transducer [2]. In addition advanced processing techniques are used to increase the resolution capabilities of ultrasonic systems.

2.1.2 Emerging Applications

Passive radio-frequency identification (RFID) transponders, i.e. transponders containing no on-board power sources, while available for many years, have been applied to humans since year 2004 [3]. These transponders are encoded and implanted in a patient, and subsequently accessed with a hand-held electromagnetic reader in a quick and non-invasive manner. Due to the small size of the transponder, the patient does not feel any discomfort or even its presence. The companion hand-held reader emits a low-frequency electromagnetic field that activates the passive transponder, which transmits its encoded data to the reader. Hence, no battery or other source of electrical power is required in the passive transponder, further reducing its size and making it even more suitable for implantation.

However, while an external RF transmitter could be used to communicate with the transponder, RF energy may only penetrate a few millimeters into a body, because of the dielectric nature of the human tissue. Therefore it may not be able to communicate effectively with a transponder that is located deep within the body. In addition, although an RF transmitter can induce a current with the transponder's receiving antenna—generally a low impedance coil—the voltage generated is too low to provide a reliable empowering mechanism. Moreover, RF telemetry systems can interfere with other RF wireless systems. In a further alternative, magnetic energy could be used to control a transponder, since a body generally does not attenuate magnetic fields. However, the presence of external magnetic fields encountered by the patient during daily activity may expose the patient to the risk of false positives, i.e. accidental activation or deactivation of the transponder. Moreover, external magnetic systems are cumbersome and are unable to effectively transfer coded information to a deep implanted transponder.

Acoustic waves overcome RF and magnetic limitations so that energy transfer and wireless communication to a deep implanted medical device or transponder is possible. In 2001 power and information transmission to an implanted medical device using ultrasound was published by Kawanabe et al. [4]. The ultrasound power link efficiency reached 20%, showing an increase of 1 °C inside the skin when a power of 1.5 W was applied to the transmitter or control unit. Moreover, dual-directional information transmission between the inside and outside of the body at different distances was demonstrated. However, misalignment has been shown to be an issue in order to have reliable communication. In 2004 an acoustically powered implantable stimulating device was patented by Penner [5], wherein an array of acoustic transducers was used to maximize the received energy into the implant and to partially solve displacement issues.

2.2   In-Vitro Platform to Study Ultrasound

A platform to study ultrasound as a source for wireless energy transfer and communication for implanted medical devices is described. A tank is used as a container for a pair of electroacoustic transducers, where a control unit is fixed to one wall of the tank and a transponder can be manually moved in three axes and rotate using a mechanical system. The tank is filled with water to allow acoustic energy and data transfer, and the system is optimized to avoid parasitic effects due to cables, reflection paths, and cross-talk problems. A printed circuit board is developed to test energy scavenging such that enough acoustic intensity is generated by the control unit to recharge a battery loaded to the transponder. In the same manner, a second printed circuit board is fabricated to study transmission of information through acoustic waves. This research work has been presented in September 2010 at the IEEE EMBC conference [6]. This work corresponds to the first implementation of the platform. An improved version, which consists of a single PCB for the transponder, will be presented in Chap. 5.

2.2.1   System Description

Figure 2.1 shows the platform block diagram, where the control unit (CU) and the transponder (TR) transducers are represented by an equivalent electrical impedance Z_{CU} and Z_{TR}, respectively. The acoustic waves are sent from the CU to the TR covering a distance d at the operating frequency f_0. A source signal, V_{IN}, is applied to the CU that generates mechanical vibrations in the direction of the TR, the output signal produced corresponds to the available voltage of the implanted circuit, V_{AV}. Following, V_{AV} is boosted up through a shunt inductor, Lp, and rectified using

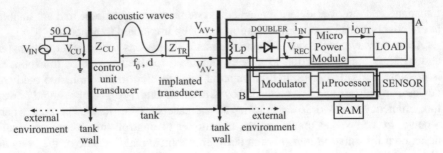

Fig. 2.1 Block diagram of the platform

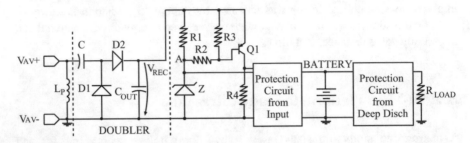

Fig. 2.2 Schematic of the front end energy harvesting circuit

an AC to DC converter. The rectified voltage, V_{REC}, has to be set at the correct constant level to recharge the battery, within the micro-power module, while keeping the delivered current, i_{IN}, high enough to obtain a quick charge. Once the energy storage element is completely charged, the load is connected to the battery that delivers the current i_{OUT}. This is the energy harvesting chain labeled as block A, while a second block B represents the wireless communication system. A simple modulator mechanism was developed to transmit a coded signal from the TR to the CU, which contains the sensor information stored in the random-access memory (RAM), through a microprocessor.

Figure 2.2 shows the complete schematic of the energy harvesting circuit, where the battery is recharged through acoustic waves. The differential output of the implanted transducer, V_{AV+} and V_{AV-}, is fed into a voltage doubler circuit which yields an output signal equal to

$$V_{REC} = 2\hat{V}_{AV+} - 2V_D \tag{2.1}$$

Where $\hat{V}_{AV+}$ is the positive peak amplitude of the implanted transducer, V_D is the threshold voltage of diodes D1, D2, and V_{AV-} is connected to the ground plane. The voltage at node A is set to 4.1 V by the reference diode. This is the voltage needed to recharge the battery when transistor Q1 is forward biased. So, the rectified voltage, V_{REC}, across the output capacitor, C_{OUT}, has to be at least 4.7 V. Considering that the active area of the receiver is quite small, this minimum required voltage is hard

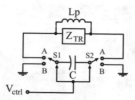

Fig. 2.3 Modulation technique: when switches S1 and S2 are in position A, a capacitance C prevents the transducer from vibrating and this results in signal reflection, thus transmitting a high state value; when switches S1 and S2 are in position B, the inductance Lp increases the magnitude of the transducer impedance, which results in signal absorption and so a low state value is transmitted

to achieve while keeping an excitation voltage, V_{CU}, low enough to prevent hot spots in the IMD. Therefore, V_{REC} can be boosted up by a factor of three by adding a high-Q parallel inductor Lp at the input of the doubler. Lp cancels out the effect of parasitics due to the transducer and the imaginary part of the input impedance of the rectifier. Also two protection circuits are used to prevent reverse current flowing toward the input and to avoid deep discharge of the battery by the load, R_{LOAD}.

Figure 2.3 shows an amplitude modulation (AM) technique known as backscattering modulation, also known as echo modulation [7]. The load seen by the implanted transducer affects the reflection or the absorption of the transmitted signal from the CU towards the TR. A μprocessor drives both switches S1 and S2 through a control signal, V_{ctrl}. If V_{ctrl} = logic-1, then S1 and S2 connect the capacitor C to Z_{TR} making the transducer stiffer and reflecting back the incoming signal so a high state is transmitted to the CU receiver. When V_{ctrl} = logic-0, then S1 and S2 short circuit C to ground and the transducer is able to vibrate allowing the incoming signal to be absorbed, thus transmitting back a low state.

2.2.2   Construction of the Platform

Figure 2.4 shows the in-vitro platform used to hold a pair of acoustic transducers, either single elements or an array, placed inside a $50 \times 50 \times 50$ cm^3 plexiglas tank. Figure 2.4a presents the pair of transducers used for the energization test made by Ferroperm Piezoceramics A/S. The CU transducer (Ferroperm Pz28) is a single element-focused transducer having a diameter of 50 mm and a radius of curvature of 50 mm. The CU transducer is attached to one of the walls of the tank, while the transponder transducer (Ferroperm Pz26) represented by a piston having a diameter of 6.35 mm and is held by a rod at a distance of 50 mm at which the maximum transfer of energy occurs. To test the acoustic wireless communication another pair of transducers are used which are made by IMASONIC, a 1 MHz flat array (64 elements) used as CU transducer and a 1 MHz flat single element transducer used for the transponder. Each element of the CU transducer has an area of 30×1.55 mm^2,

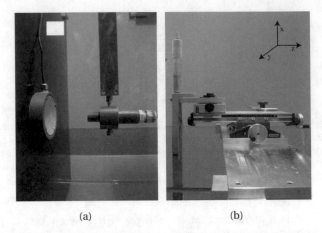

(a)   (b)

Fig. 2.4 Photograph of the in-vitro platform. (**a**) The pair of transducers used for the test, CU attached to the absorber material (white square), (**b**) mechanical system to move the TR along the x, y, and z directions

the pitch between each element is 2.05 mm, and the whole transducer bandwidth at -6 dB is 65%. The transponder transducer presents a bandwidth of 60% at -6 dB, and a diameter of 13 mm.

In order to reduce standing waves and echoes, an absorber material made of silicone is placed all around the two transducers (Aptflex F28; http://www.acoustics. co.uk/products/aptflex-f28). In Fig. 2.4a the absorber material is only present behind the CU transducer in order to clearly show the TR transducer. Figure 2.4b shows the mechanical system used to move the piston transducer inside the tank along the x, y, and z axes with a resolution of 50 μm, also having the possibility to be rotated $\pm 25°$ along the x axis.

Two different PCBs were fabricated to evaluate the ability of acoustic waves to recharge a lithium battery with a capacity of 300 μAh and a nominal output voltage of 4.1 V, manufactured by Infinity Power Solutions [8], and also to transmit information.

2.2.3 Wireless Energy Transfer

To transmit energy from the base station towards the piston, the acoustic intensity must be kept below Food and Drug Administration (FDA) standards [9]. Hence, the maximum spatial-peak temporal-average acoustic intensity is defined and corresponds to the maximum intensity occurring over the pulse repetition period (read Chap. 3 for more details). For cardiac application, the maximum derated acoustic intensity $I_{SPTA,3}$ is equal to 430 mW/cm^2, where derating is a 0.3 dB/MHz/cm factor applied to intensities.

Considering the TR piston transducer in Fig. 2.4a at a distance d from the CU, where the maximum of the transmitted acoustic intensity is detected, the received electrical power can be defined as

$$P_{E,RX} = P_{A,RX} \times \eta_{TR} \qquad (2.2)$$

where $\eta_{TR} = 10\%$ represents the electroacoustic efficiency and $P_{A,RX}$ the acoustic power at the active surface of the piston transducer which may be expressed as

$$P_{A,RX} = A_{TR} \times I_{SPTA,3} \qquad (2.3)$$

where $A_{TR} = 31.6 \times 10^{-2}$ cm^2 is the transducer aperture area. The electrical power received at the transponder loaded with a conjugate matched impedance can be written as

$$P_{E,RX} = \frac{1}{2} \times \frac{\left|2\hat{V_{AV}}\right|^2}{4R_{TR}} \qquad (2.4)$$

where R_{TR} represents the real part of the transducer impedance equal to 248 Ω at $f_0 = 1$ MHz. Combining (2.2)–(2.4), the available voltage V_{AV} can be expressed as

$$\hat{V_{AV}} = \sqrt{2 \times R_{TR} \times A_{TR} \times I_{SPTA,3} \times \eta_{TR}} \qquad (2.5)$$

The rectifier voltage V_{REC} can be estimated by substituting (2.5) in (2.1). Figure 2.5 plots V_{AV} versus the derated acoustic intensity, which is swept from 0 to 2 W/cm^2. Above the transition border set at 430 mW/cm^2 tissues may be damaged. Combining (2.1) with (2.5), the rectifier voltage can be expressed as

$$V_{REC} = 2\sqrt{2 \times R_{TR} \times A_{TR} \times I_{SPTA,3} \times \eta_{TR}} - 2V_D \qquad (2.6)$$

Fig. 2.5 Available peak voltage at the output of the piston transducer as a function of the acoustic intensity

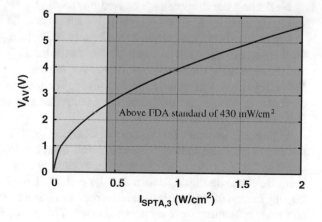

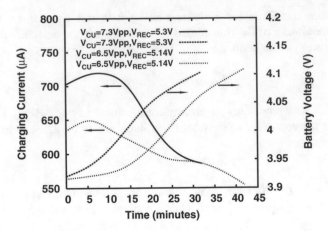

Fig. 2.6 Charging current and voltage profile

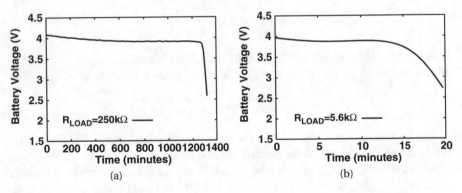

Fig. 2.7 Discharging voltage profile for different loads: (**a**) $R_{LOAD} = 250$ kΩ, (**b**) $R_{LOAD} = 5.6$ kΩ

Neglecting the drop voltage caused by the diodes (V_D), the maximum rectifier voltage that can be achieved is around 5 V at the acoustic intensity of 430 mW/cm^2.

Figure 2.6 shows the current to the battery sinks during charging and the voltage level across the battery. It is clearly visible that as the V_{CU} increases, V_{REC} increases and the recharging time decreases. Moreover, the battery is defined as charged when the voltage across its pins is equal to 4.1 V. The status of the battery charge, both current and voltage, is monitored by a voltmeter and an amperometer through a LabVIEW program installed on a PC.

Figure 2.7 presents the voltage discharging profile versus time for different resistive loads. The battery is disconnected from the load when the voltage drops down to 2.5 V by the deep discharge protection circuit to prevent the battery from being damaged. In Fig. 2.7b for a $R_{LOAD} = 5.6$ kΩ, the battery is out of charge after 20 min. In case of IMDs, the load is a sensor and it sinks current only for few seconds and then it is turned off by a μprocessor. Therefore, assuming a fully charged battery, 4.1 V, and a load of 5.6 kΩ, the delivered current is $i_{OUT} = 732$ μA. Moreover, considering an operation of the load for only 200 ms/min, the battery lasts for 4 days.

2.2.4 Wireless Communication

Figure 2.8 presents the control unit (CU) and transponder (TR) within the acoustic communication subsystem. To prove the wireless data transmission, a CU transducer made of sixty-four elements is used, where only one element is used to transmit an acoustic signal, $Z_{CU,TX}$, and an adjacent element, $Z_{CU,RX}$, is used to detect the reflected wave yielded from the TR. The distance between the control unit and the transponder is 20 cm.

Figure 2.9 shows the signal used to excite $Z_{CU,TX}$ and the received signal V_{AV}. A message is stored in the μprocessor and used as control signal, V_{ctrl}, by the backscatter modulator (Fig. 2.3).

Figure 2.10 presents V_{ctrl} and the demodulated signal, V_{dem}. The fact that V_{dem} is represented by a square wave is due to the presence of a cascade by the two amplifiers in the receiver demodulator block; a single-stage amplifier has a gain of 40 dB so that the input sinusoidal signal, V_{RX}, saturates.

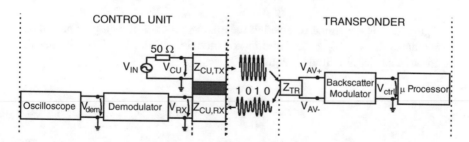

Fig. 2.8 System communication block diagram

Fig. 2.9 Transmitted signal from the control unit and received signal to the transponder

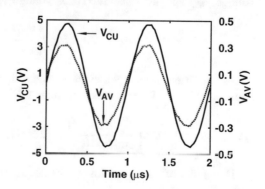

Fig. 2.10 Message transmitted from the transponder to the control unit through backscattering and demodulated signal generated by the control unit

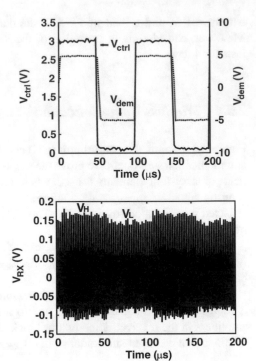

Fig. 2.11 Received signal at the control unit before being demodulated

Figure 2.11 is the signal received at the control unit before being amplified and demodulated, V_{RX}. Using the definition of modulation index in [10], the derived variation in carrier amplitude is equal to

$$MI = \frac{V_H - V_L}{V_H + V_L} = 0.077 \qquad (2.7)$$

The voltage attenuation A_V is described as the ratio between the received peak voltage V_{CU} and the transmitted peak voltage V_{RX} and is equal to

$$A_V = 20 \times log\left(\frac{V_{CU}}{V_{RX}}\right) = 28.37 \text{ dB} \qquad (2.8)$$

As the attenuation coefficient in water is equal to 0.0022 dB/cm/MHz, the attenuation due to the transmission channel is expressed as

$$A_{V,water} = 0.0022\left[\frac{\text{dB}}{\text{cm} \times \text{MHz}}\right] \times (2 \times d \text{ [cm]}) \times 1 \text{ [MHz]} = 0.088 \text{ dB} \qquad (2.9)$$

Thus most of the signal losses are due to electrical mismatches in the receiver and in the transmitter.

In the demodulator it is preferable to use a comparator with an adjustable threshold according to the minima and maxima of the received signal, V_{RX}, in order to get rid of the cross-talk between $Z_{CU,TX}$ and $Z_{CU,RX}$. Moreover, the transponder system should be autonomous when wireless communication occurs so that during testing it is necessary to avoid any external power supply, thus the battery has to be embedded.

2.3 Summary

An in-vitro platform to study ultrasound as source for wireless energy transfer and communication has been proposed. A system description of the platform, including the mechanical system to hold and move a control unit and a transponder, was given. Two PCBs were fabricated in order to be able to recharge a lithium battery and to implement a backscatter modulator. Henceforth, this set-up allows the engineer to design an in-vitro testing tool for in body acoustic wave propagation and to solve communication issues such as interference that may occur among elements of the same array, which is referred as cross-talk.

References

1. K.R. Erikson, F.J. Fry, J.P. Jones, Ultrasound in medicine-a review. IEEE Trans. Sonics Ultrason. **21**(3), 144–170 (1974)
2. J.W. Hunt, M. Arditi, F.S. Foster, Ultrasound transducers for pulse-echo medical imaging. IEEE Trans. Biomed. Eng. **30**(8), 453–481 (1983)
3. K. Michael, A. Masters, Applications of human transponder implants in mobile commerce, in *8th World Multiconference on Systemics, Cybernetics and Informatics*, vol. V, 2004, pp. 505–512
4. H. Kawanabe, T. Katane, H. Saotome, O. Saito, K. Kobayashi, Power and information transmission to implanted medical device using ultrasonic. Jpn. J. Appl. Phys. **40**(Part 1, No. 5B), 3865–3866 (2001)
5. A. Penner, Acoustically powered implantable stimulating device, 2004
6. F. Mazzilli, M. Peisino, R. Mitouassiwou, B. Cotte, P. Thoppay, C. Lafon, P. Favre, E. Meurville, C. Dehollain, In-vitro platform to study ultrasound as source for wireless energy transfer and communication for implanted medical devices, in *2010 Annual International Conference of the IEEE Engineering in Medicine and Biology Society (EMBC)*, 31 Aug–4 Sept 2010, pp. 3751–3754
7. D.A. Shoudy, G.J. Saulnier, H.A. Scarton, P.K. Das, S. Roa Prada, J.D. Ashdown, A.J. Gavens, P3F-5 an ultrasonic through-wall communication system with power harvesting, in *IEEE Ultrasonics Symposium, 2007*, Oct 2007, pp. 1848–1853
8. Infinity Power Solutions, Standard Product Selection Guide
9. Food and Drug Administration, *Information for Manufacturers Seeking Marketing Clearance of Diagnostic Ultrasound Systems and Transducers*, 2008
10. C.-S.A. Gong, M.-T. Shiue, K.-W. Yao, T.-Y. Chen, Y. Chang, C.-H. Su, A truly low-cost high-efficiency ASK demodulator based on self-sampling scheme for bioimplantable applications. IEEE Trans. Circuits Syst. I Regul. Pap. **55**(6), 1464–1477 (2008)

Chapter 3
Regulations and System Specifications

Keywords Ultrasound · Regulations · Safety limits · Acoustic pressure ·
FOCUS · Implanted medical device (IMD) · Wireless power transfer (WPT) ·
Wireless energy transfer · Food drug administration (FDA) · Wireless
communication · Energy harvesting · Health monitoring · CMOS · Transducer ·
Battery · Control unit (CU) · Attenuation · Frequency selection · Rectifier ·
Power amplifier

Ultrasound standards already exist for medical applications as imaging, diagnosis
and therapy [1, 2]. Hence, it is recommended for novel applications to make
analogies with similar systems in order to define user requirements (e.g. weight and
size of the medical device) and test procedures (e.g. in-vitro, in-vivo). In addition,
different organizations give directions for ultrasound use to avoid potential injuries
due to thermal (e.g. energy is converted into heat) and mechanical bioeffects (e.g.
cavitation). In this chapter exposure limits to ultrasound for human tissues (e.g.
cardiac) are introduced so that system specifications (e.g. operating frequency and
transducers selection) can be derived.

3.1 Safety Limits

The Food and Drug Administration (FDA) agency has established recommendations
for acoustic output from ultrasound medical devices. Therefore, the FDA sets
intensity limits for ultrasound exposure in terms of spatial-peak temporal-average
intensity (I_{SPTA}) and spatial-peak pulse-average intensity (I_{SPPA}). Before giving
the acoustic limits of I_{SPTA} and I_{SPPA}, it is required to introduce some terminology
that is going to be used in this chapter.

Figure 3.1a illustrates the effect of an ultrasound wave when it passes a point
in a medium. The particles of the medium are excited and oscillations start. The
maximum value of pressure during the passage of the pulse is the peak positive
pressure (peak compression), while the minimum value of pressure is called peak

© Springer Nature Switzerland AG 2020
F. Mazzilli, C. Dehollain, *Ultrasound Energy and Data Transfer for Medical Implants*,
Analog Circuits and Signal Processing, https://doi.org/10.1007/978-3-030-49004-1_3

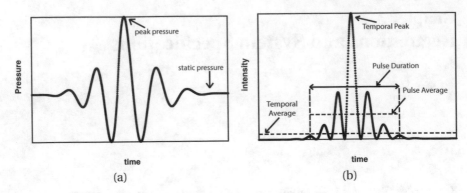

Fig. 3.1 (a) The peak positive pressure is the maximum value of pressure in the medium during the passage of an ultrasound pulse. (b) The intensity is related to the pressure squared

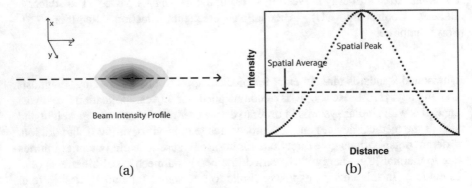

Fig. 3.2 (a) The acoustic beam intensity profile. (b) The spatial intensity profile

rarefaction. It is possible to measure the local acoustic pressure variations with the aid of a hydrophone. Figure 3.1b shows the intensity waveform derived from the pressure waveform in Fig. 3.1b. The maximum value of intensity during the passage of the pulse, within its duration (pulse duration or PD), is the temporal peak (TP) intensity. The acoustic intensity can be averaged during the pulse duration and in this case it is called pulse average (PA) or it can be averaged during the whole time frame and in this case it is called temporal average (TA). Figure 3.2a shows the beam intensity profile and Fig. 3.2b represents the intensity profile along the distance. It is possible to define the spatial peak as the point in space where the intensity is maximum and the spatial average (SA) as the average intensity along the beam profile.

Table 3.1 illustrates the limit for ultrasound exposure in terms of derated $I_{SPTA,3}$ and $I_{SPPA,3}$ [3]. All parameters have to be derated by 0.3 dB cm^{-1} MHz^{-1} to compensate for attenuation by the tissue-path.

Table 3.1 Intensity limits expressed in terms of $I_{SPTA,3}$ and $I_{SPPA,3}$ for application specific regulated by FDA

Application	$I_{SPTA,3}$ (mW/cm^2)	$I_{SPPA,3}$ (W/cm^2)
Peripheral vessel	720	190
Cardiac	430	190
Fetal, neonatal	94	190
Ophthalmic	17	28

Fig. 3.3 Mechanical index versus the operating frequency for different pressures

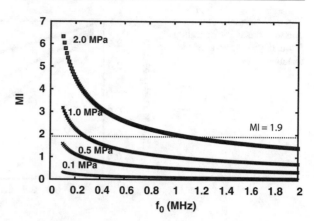

3.2   System Specifications

3.2.1   Frequency Selection

The selection of frequency for telemetry and for delivering energy to an implanted device is guided by: (1) accepted standard for diagnostic ultrasound devices and (2) physical limitations for generating an acoustic field with a single element or an array transducer. As mentioned in Sect. 3.1 FDA states that two exposure parameters must be met: I_{SPTA} for continuous wave energy delivery and I_{SPPA} for pulsed wave telemetry and energy delivery. In the latter case, the mechanical index (MI) is often specified as an estimation for the degree of bioeffects. For instance, a higher mechanical index means a larger bio-effect (cavitation), for diagnostic ultrasound the MI cannot exceed 1.9. The MI is expressed as

$$MI = \frac{p_{r,3}}{\sqrt{f_0}} \times \frac{\sqrt{1\ \text{MHz}}}{\text{MPa}} \tag{3.1}$$

where $p_{r,3}$ is the peak rarefactional pressure in MPa derated by 0.3 dB cm^{-1} MHz^{-1} and f_0 is the center frequency in MHz. To make the mechanical index unitless, the right-hand side of Eq. (3.1) is multiplied by [(1 MHz)$^{0.5}$/(1 MPa)]. Figure 3.3 shows that cavitation is more prominent at low frequencies, while higher frequencies result to be more safe.

Table 3.2 Medical frequencies

Application	Frequency	Power, intensity
Ultrasound therapeutic	1–3 MHz	15 W, 3 W/cm^2
Ultrasound diagnostic	1–40 MHz	100 W/cm^2
Ultrasound high intensity focused (HIFU)	2–6 MHz	2000 W/cm^2

Fig. 3.4 Normalized pressure versus the operating frequency as function of the distance

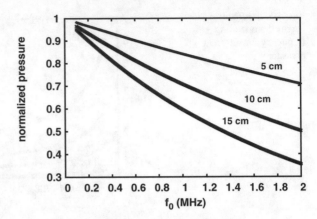

Table 3.2 gives typical frequency ranges for ultrasound medical applications; it is important to notice how low operating frequency is avoided. The main difference among the listed applications is the amount of ultrasound exposure. High intensity focused ultrasound (HIFU) is used to locally heat and destroy diseased or damaged tissue through ablation. Moreover, lower intensities are required for ultrasound therapeutic and diagnostic.

Other parameters that require attention include the attenuation factor (α)and the loss of pressure amplitude as a function of the operating frequency. The pressure can be expressed as

$$p(z) = p_0 exp^{(-\alpha \cdot z \cdot f_0)} \tag{3.2}$$

where $\alpha(f) = \alpha \cdot f_0^k$ represents the attenuation coefficient which in biological media is frequency dependent. The value of $\alpha = 0.3$ dB/cm/MHz is an average attenuation coefficient for biological soft tissues and z is the distance along the acoustic axis. The coefficient k represents a constant and in the case of tissues is assumed equal to unity [4]. Figure 3.4 shows the normalized pressure $p(z)/p_0$ versus the operating frequency as a function of the distance. The operating frequency in a high-frequency range limits the penetration of ultrasound and promote heating. To select the operating frequency a trade-off exists between undesired side effects due to cavitation or temperature elevation and penetration of the beam.

Therefore, it is necessary to define the beam width of the emitted ultrasound field from a transducer. Figure 3.5 shows a rectangular transducer where the geometrical

Fig. 3.5 Effect of element width on beam width

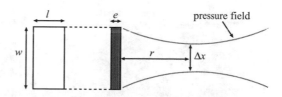

dimensions are reported as w (element width), l (length), and e (thickness). The focal point is located at distance r from the emitting source.

Hence, the beam width Δw_{beam} at 50% of the maximum intensity for a rectangular transducer can be approximated as [5]

$$\Delta w_{beam}(3\ \mathrm{dB}) \approx 0.88\frac{\lambda r}{w} \tag{3.3}$$

where λ is the acoustic wavelength expressed as

$$\lambda = \frac{v_{ac}}{f_0} \tag{3.4}$$

where v_{ac} is the acoustic velocity (1500 m s^{-1}) and f_0 is the operating frequency.

In addition, it is possible to study the beam width as a function of the operating frequency with the help of simulated results. To this end FOCUS [6] offers the possibility to simulate pressure distributions for different types of transducer shapes. A flat linear array made of 64 elements is modeled and the beam width is analyzed at the focal point (see Appendix). Figure 3.6 shows the results obtained with FOCUS at different operating frequencies along the xz plane. The focal point is located at 10 cm and as frequency increases the focal beam width decreases and the peak intensity increases. It is important to notice that at low frequency the energy is spread along the surface of the transducer while increasing the frequency causes most of energy to be concentrated at the focal point.

Figure 3.7a shows the normalized amplitude along the x axis centered at the focal point ($z = 10$ cm). As it is stated before, the focal point width is decreasing as frequency increases. Figure 3.7b represents the full-width at half-maximum of the beam width; at frequencies higher than 1 MHz the width of the focal point can be small compared to the active area of the receiving transducer.

According to the analysis done with respect to the MI, attenuation and beam width, 1 MHz is chosen as the operating frequency in order to have a good safe margin between bioeffects and depth of focus to energize a (or retrieve information from) system deep implanted into the human body.

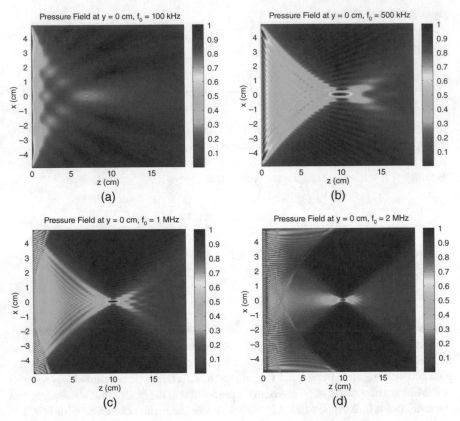

Fig. 3.6 Acoustic field simulation using FOCUS: (**a**) 100 kHz, (**b**) 500 kHz, (**c**) 1 MHz and (**d**) 2 MHz

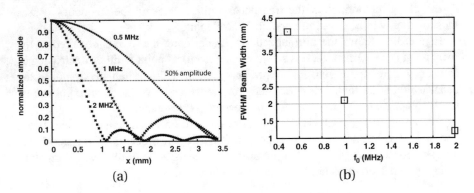

Fig. 3.7 (**a**) Focal beam profile and (**b**) full-width at half-maximum (FWHM) beam width

3.2.2 Rechargeable Battery

The main constraint in selecting the power source is the available room within the implant. The transponder may be located inside the left ventricle 80 mm × 50 mm to detect any change in its dimensions. Hence, the implant needs to occupy the smallest volume in order to not block the normal behavior of the heart; to this end, the height, width, and length are chosen to be 5 mm, 26 mm, and 13 mm, respectively. Therefore, the power source should have a small footprint and according to the type of application, the implant should remain inside the body for many years (e.g. 10 years for pacemaker). As a second requirement, the longevity of the battery turns out to be important, so parameters such as current leakage needs to be considered. Lastly, the battery is selected according to its capacity to suit the peak power consumption of the system so the number of recharging cycles per days (or per week) is also important. Table 3.3 compares different types of batteries from Cymbet Corp. and Infinite Power Solutions (IPS). The nominal voltage V_n represents the battery voltage when it is fully charged and unloaded, C represents the capacity, R_S the internal series resistance [or equivalent series resistance (ESR)], and the last column the number of recharge cycles.

IPS offers batteries with higher capacity compared to Cymbet solutions even though a large surface area is occupied. In addition, a battery with a low capacity happens to have a higher internal resistance that can be a serious drawback if a huge amount of current is suddenly requested by the system. This peak of current causes a huge voltage drop on the battery cell, so that the battery can discharge just a few minutes after the system starts operating. Other types of batteries may be considered, such as those made by Panasonic and Varta. However, as those batteries have a cylindrical shape, they tend to have a thickness higher than 1 mm. In addition, the maximum number of recharge cycles is around 1000 so their use for long term implant becomes inconvenient. As reference the reader can refer to ML-414 made by Panasonic and V-15-H made by Varta.

As the design of the electronics aimed to recharge the battery is done in a Cadence environment, it may be useful to consider a model for the battery to be used in simulations. Table 3.4 lists four models to predict the behavior of the battery in steady-state, AC, transient, and the battery runtime. However, it is demonstrated in [7] how to combine previous models as Thevenin-, Impedance-, and Runtime-Based Model, to predict accurately runtime and I-V performance of a NiMH and polymer Li-ion batteries.

Table 3.3 Low profile batteries

Chemistry	Manufact.	Type	V_n (V)	Size (mm)	C (mAh)	R_S (Ω)	# of cycles
LiPON	IPS	MEC120	4.1	25.4 12.7 0.17	0.3	110	10,000
LiPON	IPS	MEC102	4.1	25.4 50.8 0.17	2.5	22	10,000
Solid-state	Cymbet	CBC050	3.8	8 8 0.9	0.05	750	5000
Solid-state	Cymbet	CBC3105	3.8	4 5 0.9	0.005	7000	5000

Table 3.4 Comparison of various circuit models

Predicting capability	Model			
	Thevenin	Impedance	Runtime	Hybrid [7]
DC	NO	NO	YES	YES
AC	Limited	YES	NO	YES
Transient	YES	Limited	Limited	YES
Battery runtime	NO	NO	YES	YES

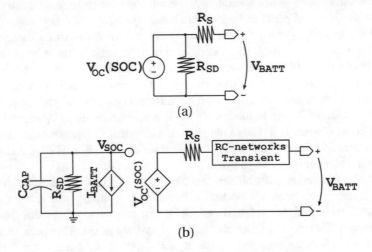

(a)

(b)

Fig. 3.8 (a) Thevenin-, and (b) hybrid-based electrical battery models

Figure 3.8 shows the Thevenin- (a) and hybrid-based electrical battery model (b). Hence, a basic form of a Thevenin-based model uses a series resistor (R_S), a voltage source represents the open-circuit voltage (V_{OC}) as function of the state-of-charge (SOC), and a parallel resistor the self-discharge resistance (R_{SD}). On the other hand, the hybrid electrical battery model uses a capacitor (C_{CAP}) and a current-controlled current source to model the capacity, SOC, and runtime of the battery. The RC networks simulate the transient response to load events and a voltage-controlled voltage source is used to link the SOC to open-circuit voltage. Moreover, the RC networks transient is made up of two different time constants and a methodology described in [8] allows an experimental evaluation via an automated test system.

The circuit in Fig. 2.2 is used to charge and discharge the MEC120 (300 μAh) and the corresponding characteristics are monitored via a software developed in LABVIEW. A GPIB interface and a data acquisition card (National Instruments USB-6009) are used to store the electrical data on the computer. Figure 3.9 shows typical current (a) and voltage (b) characteristics of the battery during the discharge phase. The MEC120 is discharged by applying a constant current ($I_{measured}$), which helps to define the capacity of the battery as

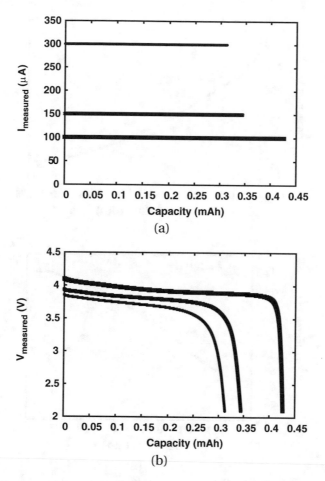

Fig. 3.9 (a) Current-, and (b) voltage-discharge characteristics of IPS MEC120 battery

$$capacity = t \times I_{measured} \qquad (3.5)$$

where t is the time that it takes to discharge the battery. It can be noticed in Fig. 3.9b that the discharge stops when the battery voltage ($V_{measured}$) reaches 2.1 V.

Figure 3.10 shows a typical current (a) voltage (b) characteristic of the battery during the charge phase. The charge stops when the battery sinks 50 μA and the battery voltage reaches 4.08 V. The battery is charged by applying a constant voltage of 5 V. But due to the current charging characteristic shown in Fig. 3.10a, it is not possible to obtain the capacity of the battery since the current average value changes versus time. However, it is possible to correlate the characteristics of the current during the charge and discharge phase by looking at the time spent to charge/discharge. Thus, the battery is charged with a constant current of 150 μA.

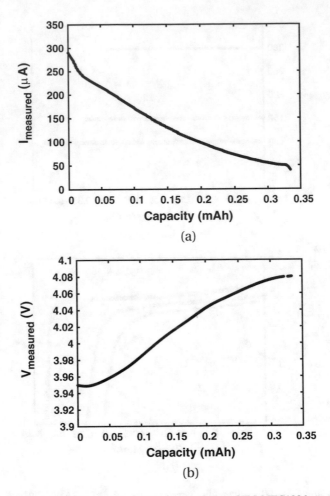

Fig. 3.10 (a) Current-, and (b) voltage-charge characteristics of IPS MEC120 battery

The $V_{OC}(SOC)$ of the battery can be determined by averaging the voltage characteristics during the charge and discharge phase that correspond to the 150 μA characteristic. Figure 3.11a shows how this methodology can be applied, where the V_{OC} curve is the average value between the charge and the discharge characteristic versus the SOC. The x-axis is obtained by normalizing the time with respect to its maximum that corresponds either to the end-of-charge (EOC) or to the end-of-discharge (EOD). Figure 3.11b shows the fitted curve of the V_{OC} obtained through a numerical method developed within the toolbox Curve Fitting provided by Matlab. The nominal battery voltage starts around 3.1 V but an IPS battery has a nominal voltage of 3.9 V. This error is due to the fact that during the charge phase, the battery needs to rest for approximately one minute before its open voltage can be measured.

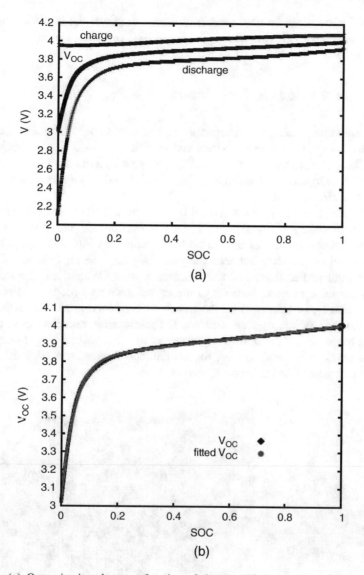

Fig. 3.11 (a) Open-circuit voltage as function of the state-of-charge $[V_{OC}(SOC)]$, and (b) comparison between $V_{OC}(SOC)$ and fitted curve using Matlab

Using the toolbox *Basic Fitting* of Matlab the curve $V_{OC}(SOC)$ can be fitted as a function of SOC. The equation below represents the $[V_{OC}(SOC)]$, where SOC corresponds to the initial condition of the battery and it is an indicator of how much the battery is initially charged.

$$V_{OC}(SOC) = 3.815e^{0.04877 \cdot SOC} - 0.7935e^{-22.11 \cdot SOC} \tag{3.6}$$

An IPS battery can be used as energy source to power the implant during the communication with the control unit and the sensing of biological activities.

3.2.3 Electroacoustic Transducers

The acoustic transducers, which serves as control unitl to transmit energy and data, is selected based on the desired shape and size of the sound beam. Figure 3.12a shows a flat circular piezoelectric crystal (or piston) with a diameter (d) of 6.35 mm that generates an unfocused beam shape, and Fig. 3.12b shows a spherical transducer array with a radius (r) of 110 mm.

The beam geometry is characterized by three zones: the near field (or Fresnel zone), the boundary (focal zone or last axial maximum), and the far field (or Fraunhofer zone). In case of unfocused transducers (see Fig. 3.12a), the location of the transition boundary between the near field and the far field is a function of the transducer frequency and the diameter crystal. Generally, if the frequency remains the same and the diameter of the crystal increases (decreases) either the far field increases (decreases) or the transition boundary is moved farther from (closer to) the transducer face. Instead, if the diameter remains equal and the resonant frequency of the ultrasound increases (decreases), the length of the near field increases (decreases). Hence, for the unfocused transducer (Fig. 3.12a), the near-field length (NFL) can be calculated as

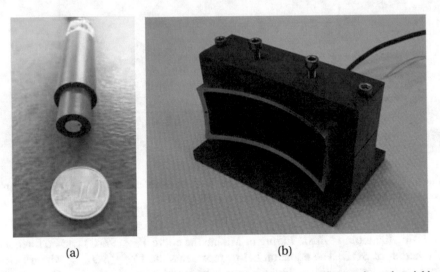

(a)   (b)

Fig. 3.12 (**a**) A flat acoustic transducer (or piston). (**b**) A convex acoustic transducer that yields a focused beam shape (or spherical transducer)

$$NFL = \frac{d^2}{4\lambda} = \frac{r^2}{\lambda} \qquad (3.7)$$

where d is the diameter, r is the radius, and λ the acoustic wavelength. For the transducer in Fig. 3.12a if the operating frequency is 1 MHz, this leads to a λ and NFL equals to 1.5 mm and 6.72 mm, respectively. However, for a focused transducer array as shown in Fig. 3.12b, because of the interactions of each of the individual beams, the formulas for a single element, unfocused transducer are not directly applicable.

The spatial peak is found by scanning over the acoustic axis z. The breakpoint distance (z_{bp}) was introduced in order to avoid measurement inaccuracies caused by measuring the ultrasound field too close to the source surface and it can be expressed as [9]

$$z_{peak} \geq z_{bp} = 1.69 \times \sqrt{A_{aprt}} \qquad (3.8)$$

where A_{aprt} is the aperture (source) surface area and it can be approximated by the area of the active element(s) of the transducer assembly. Table 3.5 gives the position of z_{bp} along the acoustic axis.

Figure 3.13 shows the pressure field measurements in the plane $(x, y) = (0, 0)$ for the piston transducer and for the spherical transducer array along the acoustic axis (z), respectively. The pressure fields are measured indirectly via the hydrophone ONDA HGL-0200, with a sensitivity ($M(f)$) of -263.3 dB (V/μPa) given at 1 MHz, which is connected to the preamplifier AH-2010 with a gain of 20 dB. The relationship that links the acoustic pressure value (p_{rms}) and the open-circuit voltage output of the hydrophone (V_{rms}) is the following:

$$P_{rms} = \frac{V_{rms}}{M(f)} \qquad (3.9)$$

The pressure field is inversely proportional to the distance in the far-field zone as shown by the Fig. 3.13a. Figure 3.13b highlights the presence of the spatial peak at 105 mm that is close to the radius of the spherical transducer. Moreover, the instantaneous acoustic intensity (I_{ac}) can be expressed as

$$I_{ac} = \frac{p_{rms}^2}{\rho v_{ac}} \qquad (3.10)$$

where ρ is the density of the propagating medium, in this case water. Moreover, the transducer yields a pressure field that it is proportional to the applied power

Table 3.5 Theoretical spatial peak

Transducer	Active area (mm^2)	z_{min} (mm)
Unfocused (Fig. 3.12a)	$3.18^2 \times \pi$	9.53
Spherical (Fig. 3.12b)	30×96	90.7

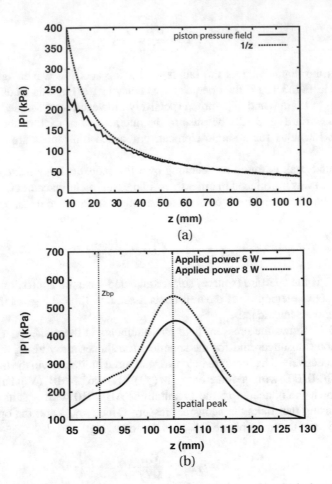

Fig. 3.13 Pressure distribution along the acoustic axis z (**a**) piston, (**b**) spherical array transducer. The data are courtesy of INSERM

(or excitation voltage). Therefore, as the excitation voltage increases the pressure field values increase. Since it is targeted to detect chronic heart failure (CHF), the implanted device is located from 8 to 12 cm from the skin surface, and the spherical transducer is preferred as control unit for obvious reasons.

Hence, the piston transducer could be a good candidate to serve as implanted crystal. However, since the diameter of the piston transducer is larger than λ, this leads to the possibility of having multiple phase excitations on the active surface, thus decreasing the amount of electrical energy that can be provided by the transducer. In addition, it is not a good solution to increase the active area of the piston because the problem of multiple phase excitation would be prominent. Therefore, it implies a small angular acceptance which can be a problem if a misalignment between the transponder and control unit is present. Figure 3.14a shows a flat transducer that is used as part of the implant. The active area is

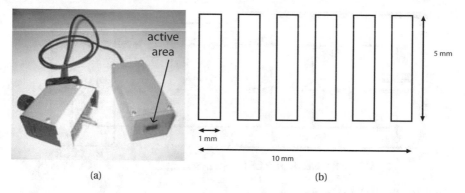

Fig. 3.14 Transponder realization: (**a**) photograph of the flat transducer array and (**b**) total active area of the transponder

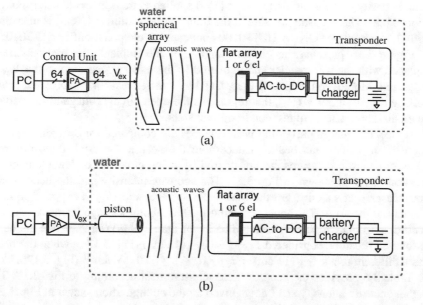

Fig. 3.15 Block diagram of the set-up used to characterize the ultrasound link in terms of FDA regulations using two different control unit (**a**) the spherical array and (**b**) the piston transducer

10 mm × 5 mm and it is divided into six elements having an area of 1 mm × 5 mm (Fig. 3.14b).

Modeling of the acoustic transducer requires finite element analysis (FEM) simulations to examine the pressure field along the different planes (xy, xz, yz). An easier method to examine the pressure field is to measure it directly with a hydrophone and correlate the peak of the pressure field (or acoustic intensity) with the excitation voltage (or applied power) to check if the FDA safety limits are met. Figure 3.15 shows the block diagram of the set-up used to link the acoustic temporal-

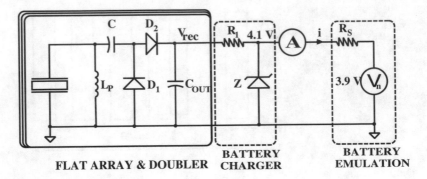

Fig. 3.16 Circuit diagram of the transponder

peak intensity to the excitation voltage (V_{ex}) applied to transducer element(s) of the control unit. The acoustic pressure at the transponder position (11 cm) is measured with the hydrophone ONDA HGL-0200 connected to the preamplifier AH-2010.

Figure 3.16 presents the circuit diagram of the implant; the hydrophone is replaced with the transponder and delivered to the battery according to the V_{ex} is measured. Moreover, the battery is replaced by a series resistor of 100 Ω (R_S) and voltage supply of 3.9 V (V_n), that represents the open-circuit voltage of the battery when discharged, to keep test conditions constant.

Hence, the energy harvesting performances are compared for different types of acoustic intensities and beams. Concerning the spherical crystal, three different types of acoustic pressures are considered by using either all or few elements of the transducer as shown in Fig. 3.17. The acoustic intensity and the beam width are measured at 11 cm from the transducer surface, and (V_{ex}) is applied to each element of the array. The acoustic spatial-peak temporal-average intensity (I_{SPTA}) according to FDA regulations should be lower than 720 mW/cm^2. When the beam is focused it can be considered $I_{SPTA} = I_{TA}(11$ cm$)$ (Fig. 3.17a); when the beam has a different focus I_{TA} (11 cm) $< I_{SPTA}(z_{bp}) \leq 720$ mW/cm ^{2} (Fig. 3.17b, c).

Table 3.6 provides the acoustic field measurement according to Fig. 3.17. The higher acoustic intensity at 11 cm is given by the configuration shown in Fig. 3.17a since most of the energy is directed at the focal point.

Concerning the piston crystal, an unfocused acoustic beam is generated while three different positions of the flat transducer array are considered as shown in Fig. 3.18. The point TR(0,0) is the origin of the coordinate system, thus $\Delta x \neq 0$ mm and $\Delta z \neq 0$ mm denote the presence of misalignment between the piston and the flat array.

Table 3.7 gives the limit for the maximum acoustic intensity at 11 cm from the piston surface, thus $I_{TA}(11$ cm$) < I_{TA}(z_{bp}) < I_{SPTA} = 720$ mW/cm^2.

The spherical transducer array and the piston are used to generate the pressure field, while the flat transducer array detects the incoming vibrations. The configurations of the CU-transponder are as follows:

Fig. 3.17 Acoustic beams generated by the spherical array: (**a**) focused beam, (**b**) focus at infinity, and (**c**) focused beam using eight central elements

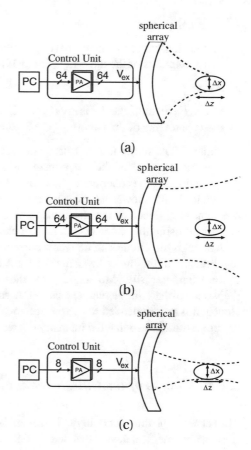

Table 3.6 Spherical transducer: limits expressed in terms of intensity and voltage

@ 11 cm	Figure 3.17a	Figure 3.17b	Figure 3.17c
$w_{beam} = \Delta x \times \Delta z$ (mm²)	2×7	88×7	17×7
I_{TA} (W/cm²)	≤ 0.72	≤ 0.54	≤ 0.61
V_{ex} (V_{pp})	≤ 1.3	≤ 5.4	≤ 5.4

Fig. 3.18 Coordinate system between the piston and the transponder (TR)

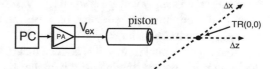

Table 3.7 Piston transducer: limits expressed in terms of intensity and voltage

@ 11 cm	$\Delta x = 0$ mm $\Delta z = 0$ mm	$\Delta x = 4.18$ mm $\Delta z = 0.37$ mm	$\Delta x = 8.23$ mm $\Delta z = 1.45$ mm
I_{TA} (W/cm²)	≤ 0.03	≤ 0.03	≤ 0.03
V_{ex} (V_{pp})	≤ 10	≤ 10	≤ 10

1. spherical array

  - one element of the flat array (Fig. 3.19a);
  - six elements of the flat array (Fig. 3.19b);

2. piston

  - one element of the flat array (Fig. 3.20a);
  - six elements of the flat array (Fig. 3.20b).

Figure 3.19a shows the measurement performed using the spherical crystal and one central element of the flat transducer. Figure 3.19b shows the measurement performed using the spherical crystal and all elements of the flat transducer.

Figure 3.20a shows the measurement performed using the piston crystal and one central element of the flat transducer. Figure 3.20b shows the measurement performed using the piston crystal and all elements of the flat transducer.

Although the circuit can deliver a current of 4 mA or more to the battery, FDA regulations safety limits allow a max of 1 mA (Fig. 3.19b). However, if a crystal with a small diameter is used to emit energy, then a maximum current of 0.7 mA can be delivered to the storage element but FDA safety limits are overcome (Fig. 3.20). Hence, it is recommended to use the spherical array as a transmitter and to increase the pulse duration so that the intensity decreases as the temporal average decreases.

3.2.4 Choice of the CMOS Technology

The choice of the technology is driven by the application. In this book, an implantable medical device that detects chronic heart failure is targeted. The IMD is going to be implanted on the heart surface, so the operational conditions of the system are changing over time. For instance, in one particular situation, it may happen that input amplitude to our system is pretty high and in another situation very low due to heart motion and orientation of the external control unit with respect to the IMD. To this end, it is recommended to avoid the use of technology with low oxide breakdown voltage ($V_{ox\text{-}break}$) even if the minimum voltage to recharge the battery is 4.1 V. For instance, the minimum $V_{ox\text{-}break}$ for a 0.18 μm technology operating at 3.3 supply voltage is equal to 7 V. Moreover, it is better to avoid working close to the minimum requirement because the transistor models are valid below $V_{ox\text{-}break}/2$ so its behavior maybe unexpected.

From experimental results, it is noticed that to transfer enough energy to recharge the lithium-ion battery, the intensity is higher than FDA regulations for continuous wave (CW) operation (see Figs. 3.19 and 3.20). FDA sets requirements to operate in CW mode as well as for pulse mode (or burst mode). The maximum intensity allowed for burst mode is referred as I_{SPPA} lower than 190 W/cm^2. So, larger amplitudes can be sent for short time duration toward the implanted device by changing the duty cycle of the voltage $V_{excitation}$. Hence, high-voltage (HV) CMOS

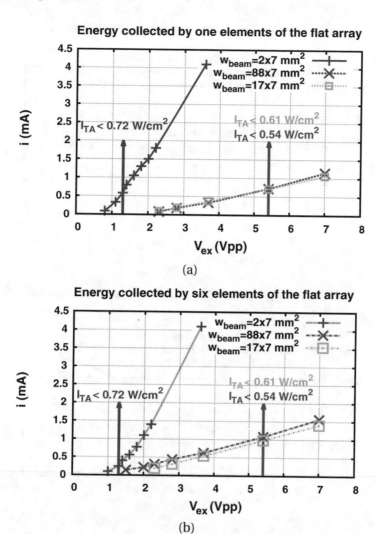

Fig. 3.19 Test performed using the spherical transducer array as control unit

technologies are needed as in case of automotive applications. XFAB XH018 technology is going to be used since HV modules are available.

Therefore, it is necessary to specify the pulse repetition period (PRP), which defines the time between transmitted pulses, to link I_{PA} and I_{TA} as

$$I_{PA} = I_{TA}\frac{PRP}{PD} \tag{3.11}$$

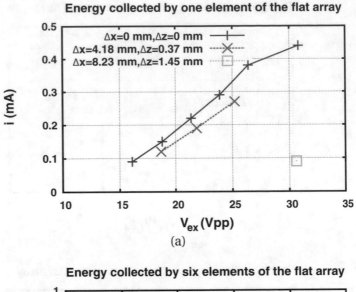

(a)

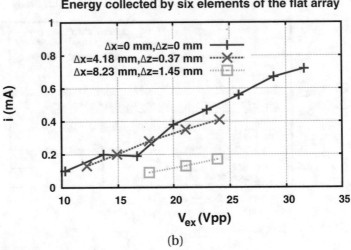

(b)

Fig. 3.20 Test performed using the piston transducer as control unit

so I_{PA} increases as PD decreases (or PRP increases). Figure 3.21 shows the graphical representations of the PRP.

If burst mode is used to transmit energy, the bandwidth (BW) of the flat transducer needs to be considered to define the minimum PD (PD_{min}). Figure 3.22 shows the bandwidth of one element of the flat transducer array (Fig. 3.14), the resonance frequency is close to 1 MHz, so the minimum pulse duration (PD_{min}) is defined as $1/BW = 2.41 \, \mu s$.

To set the maximum acoustic power that should be sent from the CU, it is important to define the maximum current and voltage that could be generated by a

Fig. 3.21 Intensity
waveform for a burst wave

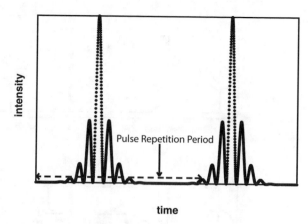

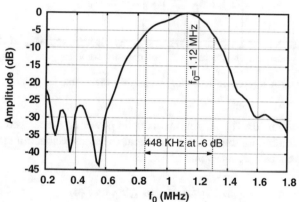

Fig. 3.22 Measured frequency spectrum of the flat array element (Fig. 3.14) in emission-reception. The graph is a courtesy of IMASONIC

single element of the flat transducer array. This maximum voltage (V_P) is influenced by the transistor oxide breakdown voltage $(V_{ox-break})$. If NMMA transistors are used to design the interface circuit, then $V_P = 2V_{ox-break} = 20$ V. Figure 3.23 shows an equivalent circuit of the electroacoustic (or acousto-electric) transducer to determine the V_P and I_P as a function of R_P [10]. In Sect. 5.1 details about the piezo equivalent model are given.

Hence, the electrical power received at resonance frequency (ω_S) can be expressed as

$$P_{E,LOAD} \leq \frac{V_P^2}{8R_S} = \frac{I_P^2 R_S Q_S^2}{8} = 41.67 \text{ mW} \tag{3.12}$$

where V_P is equal to $I_P/(\omega_S C_S)$, R_S represents the impedance of one element of the flat transducer array (≈ 1200 Ω) and $Q_S = 1/(\omega_S C_S R_S) = 1.3$ is the quality

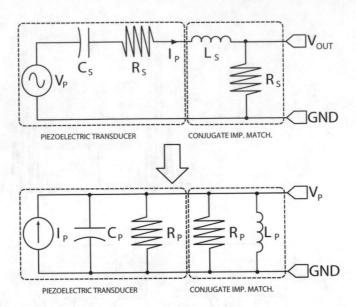

Fig. 3.23 A conjugate impedance match for maximum power extraction

factor of the transducer. The maximum acoustic power (P_A) received at the implant is derived by combining the acousto-electrical (or electroacoustic) efficiency of an element of the transducer array and (3.12) as

$$P_A \le \frac{P_{E,LOAD}}{\eta_{ae}} = 86.93 \text{ mW} \tag{3.13}$$

where the term $\eta_{ae} \approx 47.93\%$ is measured using the balance method [11] by the manufacturer. Henceforth, the maximum I_{SPPA} is defined as

$$I_{PA} \le \frac{P_A}{SA} = 1.74 \text{ W/cm}^2 \le 190 \text{ W/cm}^2 \tag{3.14}$$

where $I_{PA} = I_{SPPA}$ if the implant is located at the point of maximum intensity and SA is the active area of an element of the flat crystal (0.05 cm^2). This leads to a PRP = 5.82 μs by replacing 1.74 W/cm^2 in (3.11). Since, a charging time of 10 min is targeted a possible way to meet this requirement it is to decrease R_P so that the available current increases. This is at the cost of available voltage to the front-end circuit, as usual there is a trade-off. Using Eq. (3.12) it is possible to derive V_P and I_P as a function of R_S as

$$V_P = 2\sqrt{2 \cdot P_{E,LOAD} \cdot R_S} \tag{3.15}$$

$$I_P = 2 \sqrt{\frac{2 \cdot P_{E,LOAD}}{R_S Q_S^2}} \qquad (3.16)$$

where $P_{E,LOAD}$ is the power delivered to the interface and load circuits. Figure 3.24 shows the voltage V_P and the current I_P versus R_S at three different available power. The working area (or green area) is defined by the minimum and maximum voltage such that the transponder can function ($10 \leq V_P \leq 20$) V (or green area). For $V_P < 5$ V (or grey area) could be challenging to recharge the IPS battery as its nominal voltage is 4.1 V and for ($5 \leq V_P < 10$) V (yellow area) charge pump or boost converter circuits are required to step-up the available voltage. If V_P is higher than 20 V the transistors may be damaged.

Fig. 3.24 Voltage (**a**) and current (**b**) delivered to the interface and load circuits versus R_P

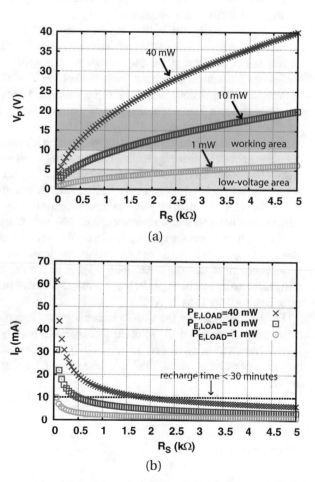

To achieve a battery recharge time (t_{charge}) between 30 and 20 min, the available current (I_P) should be at about 10 mA [12], thus the available input power $P_{E,LOAD}$ has to be higher than 10 mW with $R_S = 1200\ \Omega$.c

3.3 Summary

As described in this chapter, ultrasound standards for wireless body area networks rely on existing regulations for medical applications. Based on exposure limits to ultrasound, system specifications have been derived. An operating frequency of 1 MHz has been selected to minimize bioeffects (e.g. cavitations) and to maximize the depth of focus of the ultrasound beam.

To avoid surgery for battery replacement, a Lithium Phosphorus Oxynitride (LiPON) rechargeable battery with absolutely no possibility for thermal runaway or fire has been chosen to supply energy to the implanted device. A Thevenin model has been used to predict transient behavior of the battery when connected to the recharging circuit. Moreover, the open-circuit voltage of the battery has been characterized as a function of the state-of-charge using existing models and has been fitted using the *Matlab Basic Fitting* toolbox.

A comparison in terms of maximum emitted ultrasound pressure in water has been done between a single element flat transducer and a multiple elements spherical array. The latter has been selected to serve as a part of the control unit since it has shown higher performance while sending energy to an implanted circuit. A maximum current of 4 mA has been recovered by the implanted transducer when using the spherical transducer, and 700 µA when using the flat transducer even though maximum exposure limits to ultrasound have been not respected. The choice of the transducer to be used within the implanted device has been driven to solve problem as misalignment with respect to the control unit. To this end, a flat transducer array has been preferred to a mono element transducer. Transducers have been designed by IMASONIC.

Lastly, the choice of the CMOS technology to design the implant electronics has been pursued based on the selected rechargeable battery and previous experiments with ultrasound transducers. XFAB 0.18 µm high-voltage (HV) CMOS process has been considered since it offers thicker gate oxide thus higher oxide breakdown voltage.

In the next chapter, the control unit architecture is discussed.

References

1. International Electrotechnical Commission. Medical Electrical Equipment - Part 2–5: Particular Requirements for the Basic Safety and Essential Performance of Ultrasonic Physiotherapy Equipment, 2009
2. The British Medical Ultrasound Society. Guidelines for the safe use of diagnostic ultrasound equipment, Nov 2009

3. FDA. UCM 070911: Device Regulation Guidance
4. P.A. Narayana, J. Ophir, N.F. Maklad, The attenuation of ultrasound in biological fluids. J. Acoust. Soc. Am. **76**(1), 1–4 (1984)
5. G.S. Kino, *Acoustic Waves: Devices, Imaging, and Analog Signal Processing*. Prentice-Hall Signal Processing Series (Prentice-Hall, Englewood Cliffs, 1987)
6. Michigan State University. FOCUS. http://www.egr.msu.edu/~fultras-web/download.php
7. M. Chen, G.A.R. Mora, Accurate electrical battery model capable of predicting runtime and I-V performance. IEEE Trans. Energy Convers. **21**(2), 504–511 (2006)
8. B. Schweighofer, K.M. Raab, G. Brasseur, Modeling of high power automotive batteries by the use of an automated test system. IEEE Trans. Instrum. Meas. **52**(4), 1087–1091 (2003)
9. AIUM/NEMA. UD 2-2004 (r2009): Acoustic Output Measurement Standard for Diagnostic Ultrasound Equipment, 2004
10. K.K. Shung, M. Zippuro, Ultrasonic transducers and arrays. IEEE Eng. Med. Biol. Mag. **15**(6), 20–30 (1996)
11. S. Maruvada, G.R. Harris, B.A. Herman, Acoustic power calibration of high-intensity focused ultrasound transducer using a radiation force technique. J. Acoust. Soc. Am. **121**, 1424–1439 (2007)
12. Infinity Power Solutions. Standard Product Selection Guide

Chapter 4
System Architecture: Control Unit

Keywords Ultrasound · Control unit (CU) · Power amplifier (PA) · Operating
frequency · Attenuation · FPGA · CMOS · Acoustic pressure · Wireless power
transfer (WPT) · Wireless energy transfer · Wireless communication · Energy
harvesting · FOCUS · Beamforming · Beam steering

In the previous chapters the use of ultrasound in medicine for telemetry applications
was introduced. System specifications such as frequency, battery, and transducer
selections were given by conforming FDA regulations in terms of maximum
acoustic intensity (I_{SPPA} and I_{SPTA}). Here, the architecture of the control unit
is presented. Specifically, the circuit description addresses choices in the wireless
energy transmission and half-duplex communication. Figure 4.1 shows the block
diagram of the system architecture, where on the left side there is the control unit
(CU) and on the right side the transponder (TR). The CU is made up of the following
blocks: the power management unit (PMU), two field-programmable gate arrays
[(FPGAs) master and slave configuration], 64 power amplifiers (PAs) and high-
voltage switches, 8 low-noise amplifiers (LNAs), and the demodulation circuit. On
the other side, the TR is made up of: the energy harvesting circuit (AC-to-DC and
DC-to-DC converter, battery charger, and battery), a μprocessor, a sensor, and the
modulation/demodulation circuit.

The aim of the external control unit is to power and retrieve information directly
from the implanted transponder(s). This device should facilitate the communication
with up to about ten implants, and provide its use with data on status. An important
asset is power management during periodical charging of the implant(s). Hence, a
medical doctor or the patient needs to energize the transponder via the CU. The
charge status can be monitored continuously or periodically, and when the implant
is fully charged, the charging should terminate, with the charge time not exceeding
30 min. Reducing the charging time of the battery allows to keep the acoustic
transducer, that is linked to the control unit, for a small period of time on the
patient's body. Moreover, each element of the spherical transducer array (Fig. 3.12b)
is driven by a power amplifier that can be turned on if in transmitting mode or off if
in receiving or stand-by mode. The FPGA, that controls the PMU, is aimed to do this

© Springer Nature Switzerland AG 2020 43
F. Mazzilli, C. Dehollain, *Ultrasound Energy and Data Transfer for Medical Implants*,
Analog Circuits and Signal Processing, https://doi.org/10.1007/978-3-030-49004-1_4

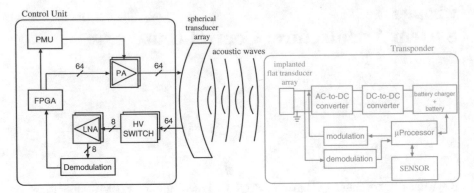

Fig. 4.1 System architecture block diagram

work. One way to communicate more efficiently with the implanted transponder(s) is beamforming, a signal processing technique used with arrays of transmitting or receiving transducers that controls the directivity of, or sensitivity to, an acoustical device. Beamforming and beam steering techniques are discussed in Sect. 4.3.

In the next section the internal blocks of the control unit are discussed with emphasis on the power management unit and the need of a master-slave configuration for the processing unit.

4.1 Front-End Architecture

The design of the control unit (CU) can be divided in two parts: (1) the front-end system that interfaces the crystals (e.g. power amplifier, low-noise amplifier, processor, etc.) and (2) the interface with the external world [e.g. Bluetooth, ZigBee, local area network (LAN)]. To properly interface the transducers, the choice of the internal blocks of the CU depends on the type of coupling employed for wireless energy harvesting/communication [e.g. electromagnetic (far field), inductive (near field), ultrasound], the number of transmitting/receiving elements, and the processor unit which is often FPGA-based. While, the design of the control unit, which is used to interface the external world is quite standard and can be developed with available commercial products.

The location of the sensor node determines the type of coupling that is more favorable to transmit energy and retrieve the sensor data. Electromagnetic coupling can be used for wearable computing devices [1, 2], inductive coupling for transcutaneous implanted devices [3], and ultrasound coupling for devices deeply implanted in the human body [4–6]. Energy transmission for magnetic-based coupling can be performed in the high-frequency (HF) spectrum since sufficient to penetrate the tissue, while limiting the data rate. Hence, dual-band telemetry systems have been developed to meet both high data rate requirement and high amount of power to

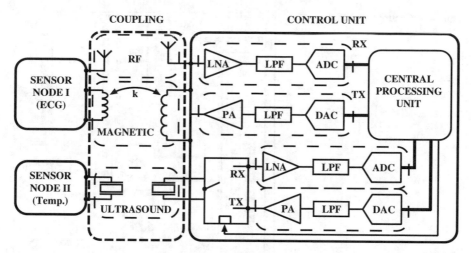

Fig. 4.2 Architecture of a general body sensor network

efficiently energize the sensor node [7]. Recently, a hybrid interface for deeply implantable medical devices has been developed, which exploits both magnetic and ultrasound advantages [8]. However, the intermediate energy conversion from magnetic to acoustic adds complexity and costs to the whole system. A serious issue in remote powering within the near-field is due to misalignment between the control unit and the implanted device. To solve the problem of misalignment multi-coil systems have been envisaged so that the wireless power transfer can be reliable [9, 10].

Figure 4.2 shows a block diagram of a sensor network, where common coupling sources are presented. Three sub-blocks can be distinguished in the CU: the transmitter (TX), the receiver (RX), and the signal processing unit. To transmit energy and data as the identification number (ID) to the sensor node, the transmitter chain is made up of a power amplifier (PA), a low-pass filter (LPF), and a digital-to-analog converter (DAC). To retrieve the data collected by the sensor node, the receiver chain is made up of a low-noise amplifier (LNA), a low-pass filter, and an analog-to-digital converter (ADC). The synchronization among the different operations is controlled by the central processing unit (CPU). For instance in case the sensor node I is addressed, energy transfer and communication can be performed at the same time, while in case the sensor II is addressed the use of a high-voltage switch is recommended to select the operation to perform either transmission or reception.

In our application, a deep implanted sensor node has to communicate with the control unit. In such a case, an electroacoustic array is used both for energy transfer and data communication.

Figure 4.3 shows the architecture block diagram of the control unit. The crystals are connected to the board via the Hypertac mini modular rectangular connector H series plus one meter coaxial cable. To allow flexibility each element of the array can

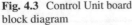

Fig. 4.3 Control Unit board
block diagram

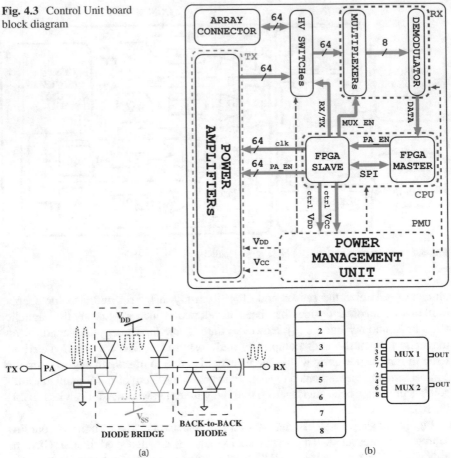

Fig. 4.4 (**a**) Block diagram of the MAX4937. (**b**) Simplified block diagram of the multiplexers chain

be used either in transmission or reception mode. It is recalled that the transducer is an array constituted by 64-elements (Sect. 3.2.3). To this end the MAX4937 octal high-voltage switch is employed, offering −75 dB isolation from input (crystal) to low-voltage output (LNA) when in off-mode and −71 dB crosstalk between any two input channels (crystals). Moreover, the input to the LNA is limited to ±1.5 V, this sets an important specification for the transmitted signal amplitude from the sensor node. To limit the number of multiplexers so as to decrease the complexity of the system, eight ADG1608 with eight inputs and one output each are used in the receiver path; thus eight channels can be selected in the demodulation process.

Figure 4.4a shows the block diagram of the MAX4937 high-voltage switch; the back-to-back diodes limit the output amplitude to ±1.5 V when the diode bridge is activated.

Figure 4.4b shows an example of the architecture for the selection process of the receiving channels. Eight elements of the array are connected to two different multiplexers (4-to-1) in order to be able to select two adjacent channels.

The slave FPGA of central processing unit enables/disables the HV switches thus setting the piezo in reception/transmission mode via the signal RX/TX and sets the multiplexers output to the wanted receiving channels via the enable signal MUX_EN.

The design of the CPU is based on a master-slave configuration in order to divide the different tasks as setting of supply voltages for the power amplifiers, transmission of ID to the implanted devices, demodulation of the sensor data, and to avoid possible short circuits between the PA output and LNA input. Hence, two Cyclone III FPGA modules, EP3C16 (slave) and EP3C40 (master), are used to this end.

4.1.1 Master and Slave Architecture

To facilitate the operations in the control unit and to have sufficient general purpose I/O, the central processing unit is based on a master-slave architecture. The master FPGA controls peripherals such as memory (8 MB flash, 2 KB EEPROM) to expand the internal capacity, USB to connect to an external PC, recovers the demodulator data and integrates a NIOS μcontroller to store the control unit firmware. On the other side, the slave FPGA enables/disables the power amplifiers, generates the gate signals, and controls the HV switch as well as the multiplexers. A 25 MHz quartz oscillator, Xpresso HC73, is used as a reference to generate all clocks available on the board via phase-locked-loops (PLL) available on the FPGAs. Hence, the master FPGA provides a 60 MHz clock frequency for the SPI interface and a 20 MHz signal reference for the slave FPGA. To allow a minimum phase shift of 3° among the gate signals of the different power amplifiers, the 20 MHz signal reference is upconverted to 120 MHz by the slave FPGA. In this manner to obtain the working frequency of 1 MHz, a frequency division technique is implemented with a counter. By delaying the counting process, a phase shift is achieved.

At the start-up of the board, the firmware configures the elements of the array either in transmission or in reception, enabling or disabling the corresponding HV switches and multiplexers. To prevent over heating or damaging of low-voltage electronics (e.g. LNAs), the slave FPGA implements a hardware protection that forbids to turn on the HV switch in case the corresponding PA is already enabled, and vice versa.

The selection of the power amplifier used to drive the acoustic transducer is discussed next.

4.2   Class-E Power Amplifier

A power amplifier is a key block in energy scavenging applications since it drives a transducer (e.g. coil, antenna, crystal) to transmit high level of energy towards the harvester. A good candidate in energy transmission is the class-E type switching power amplifier [11] with a theoretical drain efficiency up to 100% compared to linear PA (e.g. A, B, and AB) which exhibits a maximum efficiency of 50%.

Figure 4.5 shows the shunt-C class-E power amplifier along with the driving amplifier and a narrow band transducer equivalent model represented by a parallel branch constituted by the capacitor (C_t) in series with the resistor (R_t). This model is valid in the vicinity of the resonant frequency of the transducer.

The Zero-Voltage-Switching (ZVS) condition is highlighted in the dotted box, wherein the voltage (V_{DRAIN}) and the current (I_{DS}) at the drain of the transistor are depicted. The maximum drain voltage that can be achieved is 3.66 times the supply voltage (V_{CC}) [12]. While the transistor is conducting, I_{DS} is above zero and V_{DRAIN} should stay equal to zero. If this condition does not hold, then ZVS condition is not achieved and two different cases are possible: above ZVS or below ZVS.

The parallel inductor L_P resonates with the transducer at 1 MHz, the series capacitor C_S prevents any DC feedthrough and, the parallel capacitors C_{SHUNT} and C_P help to achieve zero voltage switching (ZVS). The series inductor L_S increases the power amplifier efficiency since the AC current through L_P decreases and as a consequence the iron loss decreases.

4.2.1   Tuning Methodology

Design equations have been developed for the class-E power amplifier [13], henceforth a tuning strategy is proposed. The shunt-C class-E power amplifier is

Fig. 4.5 Shunt-C class-E power amplifier with transducer equivalent model

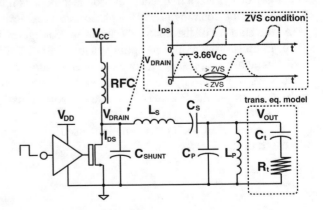

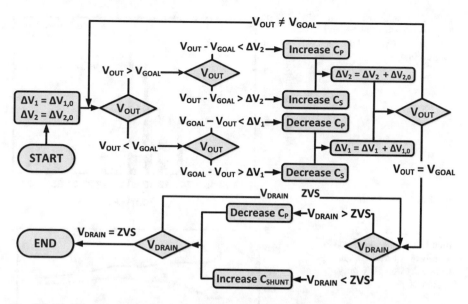

Fig. 4.6 Tuning methodology for the calibration of multiple shunt-C class-E power amplifier

designed with Advanced Design System (ADS) and an average value for R_t and C_t is considered. 64-PA are fabricated on standard substrate FR4 due to the low-frequency requirement.

Figure 4.6 shows the tuning methodology adopted to calibrate 64-PAs, where two main loops can be distinguished to set respectively the output voltage (V_{OUT}) of the PA to the desired voltage (V_{GOAL}) and to achieve ZVS for best efficiency. To set the value for the capacitors the magnitude $|V_{OUT} - V_{GOAL}|$ is compared with ΔV_1 or ΔV_2. The initial conditions $\Delta V_{1,0}$, $\Delta V_{2,0}$ for ΔV_1, ΔV_2 are chosen between 0.1 and 1 V and their values also determine the final maximum distance between V_{OUT} and V_{GOAL}.

4.2.2   Measurements

Figure 4.7 represents the histogram of the number of amplifiers versus the measured output voltage before and after using the tuning methodology. 23 PAs over 64 reach $V_{OUT} = V_{GOAL} = 19$ Vpeak, while the remaining 41 PAs present a shift only of ± 1 V from V_{GOAL}.

To evaluate the efficiency of the shunt-C class-E power amplifier a prototype is realized and the transducer is replaced by its series lumped equivalent model ($R_t - 120\ \Omega$, $X_t = -j290\ \Omega$). Figure 4.8 shows the prototype efficiency versus the DC power supply (V_{CC}) while V_{DD} is kept constant (4.5 V). Drain efficiency

Fig. 4.7 Number of amplifiers versus the output voltage with (w/) and without (w/o) tuning

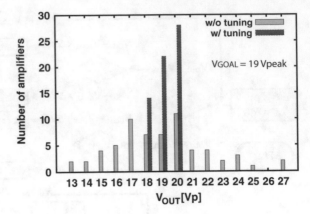

Fig. 4.8 Efficiency of the tuned shunt-C class-E power amplifier at 1 MHz operating frequency with 50% duty cycle (DC)

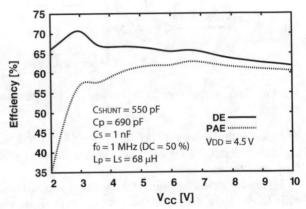

(DE, Eq. (4.1a)) and power added efficiency (PAE, Eq. (4.1b)) are measured for the power amplifier.

$$DE = \frac{P_{OUT,RF}}{P_{V_{CC}}} \tag{4.1a}$$

$$PAE = \frac{P_{OUT,RF} - P_{DYN}}{P_{V_{CC}}} \tag{4.1b}$$

where $P_{OUT,RF}$ is the delivered power to R_t, $P_{V_{CC}}$ is the DC power, and P_{DYN} is the dynamic power consumed by the driver during the charge/discharge of the gate-to-source and gate-to-drain capacitances of the power MOS. According to Fig. 4.8, the ripple on DE is less that 10% on all the scale of V_{CC} whereas the ripple on PAE is less that 10% for V_{CC} larger than 3 V.

However, the PA is subjected to ohmic losses since each passive component shows a low equivalent series resistance as well as the on-resistance of the power transistor is not zero. Moreover, the inductors effect the PA efficiency due to their ferromagnetic core which dissipates part of the stored energy. The circuit in Fig. 4.5

Fig. 4.9 Efficiency of the
tuned shunt-C class-E power
amplifier versus the power
supply V_{CC} of the buffer

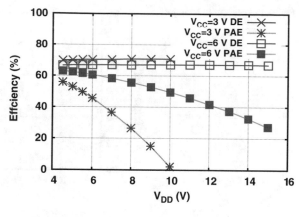

Fig. 4.10 Efficiency of the
tuned shunt-C class-E power
amplifier versus operating
frequency

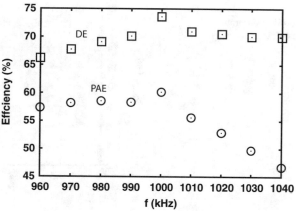

is simulated using the value reported in Fig. 4.8 and keeping V_{CC} equal to 3 V. The simulated DE results equal to 93.8% without including any loss, while subtracting the losses due to L_P and L_S, the efficiency drops down to 84.4%.

Figure 4.9 shows the power added efficiency and the drain efficiency versus V_{DD} while keeping V_{CC} constant. For $V_{CC} = 3$ V and 6 V, the DE remains constant around 70% and 67%, respectively. For $V_{CC} = 3$ V, the PAE decreases drastically when V_{DD} is twice V_{CC}; while for $V_{CC} = 6$ V, the PAE decreases linearly as V_{DD} increases and reaches a minimum of 25% at V_{DD} equal to 15 V.

Figure 4.10 shows the PA efficiency on the y-axis versus the operating frequency on the x-axis. The maximum efficiency is found at the resonance frequency (1 MHz), while out of this value the DE efficiency and the PAE efficiency respectively drop by 10% and by 15%.

Figure 4.11 shows the measured average DE and PAE of 64-PAs versus the supply voltage V_{CC}. A 73.43% drain efficiency is reached at minimum supply voltage, and a 62.54% at maximum supply voltage.

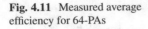

Fig. 4.11 Measured average
efficiency for 64-PAs

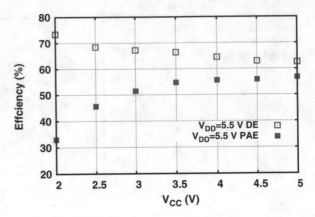

Fig. 4.12 Measured DC
power consumption for
64-PAs

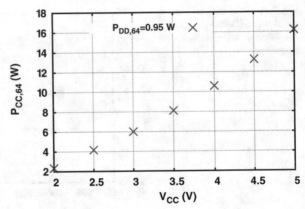

Table 4.1 Different configurations of the 64-PAs

# elements	V_{CC} (V)	Power (W)	Configuration	I_{MAX} (W/cm^2)
32	3	3	Center elements	5
32	3	3	Sides elements	4
48	3	4.5	Center elements	10.5
64	3	6	Focused	14
64	3.5	8	Focused	23
40	3	3.5	Unfocused	0.36

Figure 4.12 represents the DC power consumption due to 64-PAs versus the supply voltage V_{CC} while keeping the driver power consumption constant to 0.95 W with $V_{DD} = 5.5$ V.

Table 4.1 gives the recorded maximum acoustic intensity at 10.5 cm from the CU via the hydrophone HGL-0200. Six different configurations are used varying the number of employed power amplifiers to characterize the power consumption and emitted acoustic intensity.

The maximum acoustic intensity is recorded using 64 power amplifiers, thus increasing the power consumption. The reported power in Watts is the electrical power consumed by the transducers as a function of the supply voltage V_{CC}.

4.3 Beamforming and Beam Steering Techniques

Beamforming refers to the process of adjusting the delays in electrical signals between the individual elements of an array to focus the acoustic beam at a given depth and location. To achieve this, the signals must be adjusted both during transmission and when processing the received signals. The highest resolution for an ultrasound system is obtained when all the elements are focused at a single point in the imaging field. On the other side, a beam steering technique changes the direction of the main lobe in the radiation pattern. These two techniques can be combined for instance in space-division multiple access (SDMA) wireless communication networks or in ultrasound to elongate the depth of focus of transmit beams [14–16]; especially more work has been done in the context of linear phased array for ultrasound propagation [17–20].

To optimize the energy transmission phase towards the implant, the control unit can scan automatically the body and mark the position of the implanted device(s). When a signal is received from the transponder back to the control unit, beamforming can increase the receiver sensitivity in the direction of wanted signals and decrease the sensitivity in the direction where the signal has to be sent.

Concepts of beamforming and beam steering techniques are mentioned but out of the topic of this book. However, the hardware developed in this chapter can be used to perform such techniques by programming the FPGA that controls the power amplifier(s) of the element(s) of the array.

The spherical acoustic transducer is modeled with the help of a free software to show the operating principle of beam focusing. Wireless body sensor networks, where up to ten implants are used, can benefit from the beam focusing technique. The spherical array (Fig. 3.12b) is modeled using FOCUS [21] and the pressure field is analyzed with and without applying the beamforming technique. To prove the strength of the simulated results, the pressure distribution shown in Fig. 3.13b is normalized to its peak pressure and compared with the results obtained with FOCUS. A good agreement between the measured and the simulated results, thus FOCUS can be assumed a sufficient proof to demonstrate the beamforming technique (Fig. 4.13b).

Figure 4.14 shows a new position of the original focal point from $[xyz] = [0\,0\,10.5]$ cm to $[xyz] = [20\,10.5]$ cm. This system can be used to track with good precision the motion of the implant if located on the heart surface, whose shape changes due to respiration [22].

An implant must be localized before being tracked, Fig. 4.15 shows a possible pressure field distribution either to map a certain body area (e.g. sonogram) or to spread enough energy such that an implant can be energized and send back its position to the control unit.

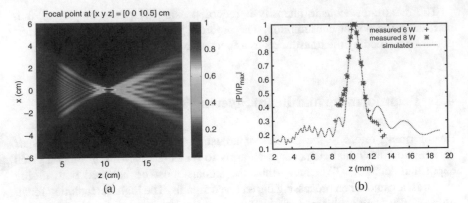

(a)   (b)

Fig. 4.13 (a) Map of the pressure field with a focal point located at $[xyz] = [0\ 0\ 10.5]$ cm. (b) Normalized pressure along the z axis with $[xy] = [0\ 0]$ cm and comparison with measured data obtained in Fig. 3.13b

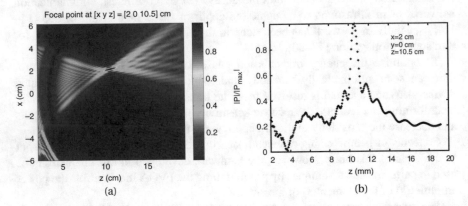

(a)   (b)

Fig. 4.14 (a) Map of the pressure field with a focal point located at $[xyz] = [2010.5]$ cm. (b) Normalized pressure along the z axis with $[xy] = [20]$ cm

The user of this type of application could imagine the different situations where the implant is first energized, then localized so that the data transfer can start.

In the next section, the demodulator architecture is discussed.

4.4   Receiver

As the control unit is used to receive the echo sent by the transponder, eight elements of the spherical transducer array are used in reception and the corresponding PAs are turned off via the FPGA (see Fig. 4.1). The eight channels are chosen manually by the user according to the test that needs to be performed. However, an algorithm may be implemented in the FPGA in order to dynamically select the eight channel

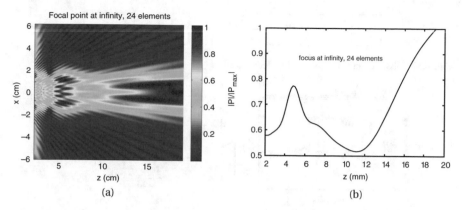

Fig. 4.15 (**a**) Map of the pressure field with a focal point located at infinity by using 24 elements of the spherical array. (**b**) Normalized pressure along the z axis with $[xy] = [00]$ cm

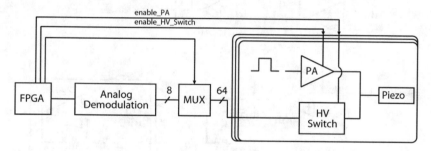

Fig. 4.16 Block diagram of the control unit in receiving mode

to increase the receiver sensitivity and selectivity. Figure 4.16 depicts the block diagram of the system used to receive any signal sent from the transponder. The FPGA enables the HV switch and disables the power amplifier corresponding to the element employed in reception. In this manner, the signal yields by the transducer are connected via the HV switch to the multiplexer (MUX). To prevent the presence of a short between the PA and the MUX, in the main of the FPGA code, at the start-up of the CU all power amplifiers are turned off and before any PA is enabled the corresponding HV switch is checked and it is verified that is enabled.

Figure 4.17 shows the block diagram of the analog demodulation. First, the incoming signal from the transponder is amplified and immediately rectified. As eight channels are employed in reception, the rectified signals are added, in this way the receiver sensitivity increases and destructive interference is avoided. Then, the envelope of the sum is detected and a low-pass-filter (LPF) is used to reject noise at high frequency mostly due to the reverberations in the transmission media (e.g. skin, bones, liver, etc.). Lastly, a series of gain stages are used to decode the incoming information sent by the transponder. The type of modulation implemented in the implanted device is discussed in Sect. 5.7; as a preview a load modulation technique is used as in RF communication.

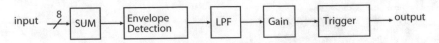

Fig. 4.17 Block diagram of the analog demodulation

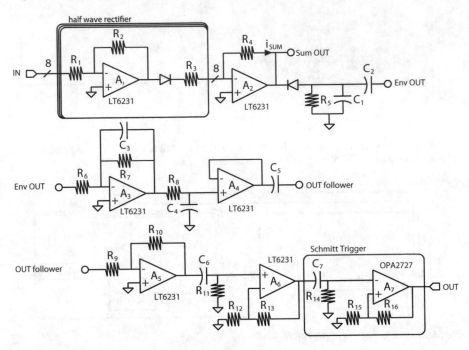

Fig. 4.18 Circuit diagram of the analog demodulation

Figure 4.18 shows the circuit diagram of the analog demodulation. A low-noise-amplifier (A_1) is connected to the transducer element via the HV switch and the MUX, and the resistor ratio $R_2/R_1 = 100k/20k$ determines its gain (G). The diode serves as rectifier to block the negative half of the incoming signal; another amplifier (A_2) is used to provide additional gain $R_4/R_3 = 90k/10k$. A second-order low-pass filter is made by the amplifier A_3 and A_4; its cut-off frequency is about 20 kHz, so that the carrier frequency at 1 MHz is suppressed. Lastly, two amplifiers, A_5 and A_6, add 60 dB gain, $R_{10}/R_9 = 100k/5k$, and $R_{13}/R_{12} = 100k/2.5k$, before that the last amplifier (A_7) detects the signal.

The demodulator is simulated in ADS to check qualitatively its behavior; since a model for the ultrasound link has not been developed, the reflected signal from the transponder is emulated via the modulator *AM ModTuned*, ready by default in ADS. Figure 4.19 sketches the generation of a single-end signal for ASK modulation, the node *IN* (RF_{OUT}) is directly connected to the negative terminal of the amplifier A_1. Hence, eight modules as in Fig. 4.19 are added to the simulation to mimic the eight channels of the receiver. The carrier frequency is 1 MHz and the data rate is 10 kbps

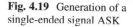

Fig. 4.19 Generation of a single-ended signal ASK

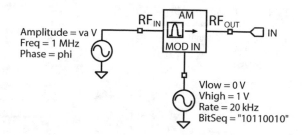

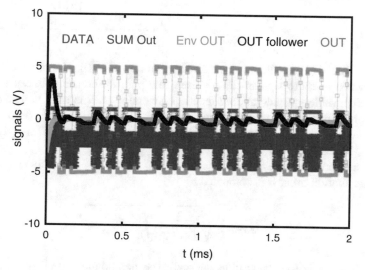

Fig. 4.20 Simulation result of the direct conversion receiver

since Manchester coding is used. Moreover, the phase (*phi*) and the amplitude (*va*) of the signal RF_{IN} are chosen following a random uniform phase distribution and avoiding to saturate the receiver, respectively.

Therefore, the receiver is simulated in order to show the advantage to have either one or more channel(s) to demodulate the data signal. A transient simulation is performed by keeping the modulation index (MI) equal to 100%. It is also assumed that the signal is arriving at time equal to zero to each channel so phase difference is not present (*phi* = 0) and the signal amplitude is equal to 1 V. The sensitivity and the bit error rate (BER), both indicators of the receiver quality, are not simulated. This is due to the limitations raised by the simulator and by the accuracy of the results. Figure 4.20 shows the waveforms of the direct conversion receiver; it can be assumed that about 500 μs is required before the signal *Env OUT* settles and follows any change in the signal *SUM Out*. This is due to the large time constant of the envelope detector given by $R_5 \times C_1 = 5 \text{ k}\Omega \times 100 \text{ nF} = 500 \text{ μs}$.

Table 4.2 compares the envelope root-mean-square (rms) value $i_{SUM,rms}$ according to the number of channels employed during the demodulation. The root-mean-square increases as the number of channels used in the demodulation process increases. Two different situations are considered: (1) all signals are in phase

Table 4.2 Comparison in terms of $i_{SUM,rms}$ according to the number of channels employed in the demodulation

# Used channels	$i_{SUM,rms}$ (μA)	
	$\Delta phi = 0$	$\Delta phi \neq 0$
1	30.7	30.7
2	43.9	44.1
3	47.7	47.9
4	50.2	52.3
5	51.9	55.8
6	53.3	56.6
7	54.12	57.3
8	54.12	57.8

($\Delta phi = 0$) and (2) a uniform phase distribution ($\Delta phi \neq 0$). The phase difference between two or more signals can represent the path difference between traveling waves. The Matlab pseudorandom function generator *rand* is used to generate the phase for eight different channels, and $\Delta phi \neq 0$ means that a non-zero delay is present between signals. Therefore, the phases are set to $0°$, $7.8°$, $27.34°$, $117.54°$, $171.66°$, $184.99°$, $261.07°$, and $316.82°$.

The envelope root-mean-square $i_{SUM,rms}$ for number of channels employed in the demodulation process is computed as

$$i_{SUM,rms} = \sqrt{\frac{1}{T_2 - T_1} \int_{T_1}^{T_2} i_{SUM}^2 dt} \tag{4.2}$$

where $T_2 - T_1$ is the integration interval. The demodulator performance is not affected by phase differences that occur among the incoming signals.

Figure 4.21 shows a picture of the control unit where four vertical PCBs on the left represent the 64 PAs, the huge cylindrical capacitors are used for the PMU of the whole board, a small vertical board on the right is the analog demodulator, and in the center two FPGAs are used to perform the system operations as sending energy to and retrieve data from the transponder.

4.5   Summary

The design of the control unit that drives the array of 64-elements has been presented in this chapter. To facilitate the operations as energy transfer and half-duplex communication, the central processing unit has been based on a master-slave architecture, implemented with two field-programmable gate arrays (FPGAs).

To send energy and data toward a deep implanted medical device, a class-E power amplifier has been designed to drive the array. Drain efficiency (DE) and power added efficiency (PAE) have been measured for different supply conditions. A DE

Fig. 4.21 Control unit board

and PAE of 70% and 57% have been measured when the supply voltage is 3 V. An acoustic intensity of 23 W/cm^2 at 10.5 cm has been recorded in water when the ultrasound transducer has been connected to the power amplifiers. The electrical power consumption due to the amplifiers was equal to 8 W.

An eight channels receiver has been designed to retrieve the sense data from the implant. Simulation results have been presented while measurements will be shown in Chap. 6.

References

1. J. Yoo, L. Yan, S. Lee, Y. Kim, H.-J. Yoo, A 5.2 mw self-configured wearable body sensor network controller and a 12 μ W wirelessly powered sensor for a continuous health monitoring system. IEEE J. Solid-State Circuits **45**(1), 178–188 (2010)
2. J. Cheng, L. Xia, C. Ma, Y. Lian, X. Xu, C.P. Yue, Z. Hong, P.Y. Chiang, A near-threshold, multi-node, wireless body area sensor network powered by RF energy harvesting, in *2012 IEEE Custom Integrated Circuits Conference (CICC)*, Sept 2012, pp. 1–4
3. C.M. Zierhofer, E.S. Hochmair, High-efficiency coupling-insensitive transcutaneous power and data transmission via an inductive link. IEEE Trans. Biomed. Eng. **37**(7), 716–722 (1990)
4. S. Arra, J. Leskinen, J. Heikkila, J. Vanhala, Ultrasonic power and data link for wireless

implantable applications, in *2nd International Symposium on Wireless Pervasive Computing, 2007, ISWPC '07*, Feb 2007

5. S. Ozeri, D. Shmilovitz, Ultrasonic transcutaneous energy transfer for powering implanted devices. Ultrasonics **50**(6), 556–566 (2010)
6. T. Maleki, N. Cao, S.H. Song, C. Kao, S.-C. Ko, B. Ziaie, An ultrasonically powered implantable micro-oxygen generator (IMOG). IEEE Trans. Biomed. Eng. **58**(11), 3104–3111 (2011)
7. G. Wang, W. Liu, M. Sivaprakasam, G.A. Kendir, Design and analysis of an adaptive transcutaneous power telemetry for biomedical implants. IEEE Trans. Circuits Syst. I Regul. Pap. **52**(10), 2109–2117 (2005)
8. A. Sanni, A. Vilches, C. Toumazou, Inductive and ultrasonic multi-tier interface for low-power, deeply implantable medical devices. IEEE Trans. Biomed. Circuits Syst. **6**(4), 297–308 (2012)
9. R. Carta, M. Sfakiotakis, N. Pateromichelakis, J. Thone, D.P. Tsakiris, R. Puers, A multi-coil inductive powering system for an endoscopic capsule with vibratory actuation. Sens. Actuators A Phys. **172**(1), 253–258 (2011)
10. E.G. Kilinc, B. Canovas, F. Maloberti, C. Dehollain, Intelligent cage for remotely powered freely moving animal telemetry systems, in *2012 IEEE International Symposium on Circuits and Systems (ISCAS)*, May 2012, pp. 2207–2210
11. P. Reynaert, M. Steyaert, *RF Power Amplifiers for Mobile Communications* (Springer, New York, 2006)
12. T.H. Lee, *The Design of CMOS Radio-Frequency Integrated Circuits* (Cambridge University Press, Cambridge, 2004)
13. M. Acar, A.J. Annema, B. Nauta, Analytical design equations for class-E power amplifiers. IEEE Trans. Circuits Syst. I Regul. Pap. **54**(12), 2706–2717 (2007)
14. F. Hooi, K. Thomenius, R. Fisher, P. Carson, Hybrid beamforming and steering with reconfigurable arrays. IEEE Trans. Ultrason. Ferroelectr. Freq. Control **57**(6), 1311–1319 (2010)
15. H. Liu, S. Gao, T.H. Loh, Compact dual-band antenna with electronic beam-steering and beamforming capability. IEEE Antennas Wirel. Propag. Lett. **10**, 1349–1352 (2011)
16. J. Atkins, Robust beamforming and steering of arbitrary beam patterns using spherical arrays, in *2011 IEEE Workshop on Applications of Signal Processing to Audio and Acoustics (WASPAA)*, Oct 2011, pp. 237–240
17. D.H. Turnbull, F.S. Foster, Beam steering with pulsed two-dimensional transducer arrays. IEEE Trans. Ultrason. Ferroelectr. Freq. Control **38**(4), 320–333 (1991)
18. M. Karaman, A. Atalar, H. Koymen, VLSI circuits for adaptive digital beamforming in ultrasound imaging. IEEE Trans. Med. Imaging **12**(4), 711–720 (1993)
19. L. Azar, Y. Shi, S.-C. Wooh, Beam focusing behavior of linear phased arrays. NDT E Int. **33**(3), 189–198 (2000)
20. M.A. Hassan, A.M. Youssef, Y.M. Kadah, Modular FPGA-based digital ultrasound beamforming, in *2011 1st Middle East Conference on Biomedical Engineering (MECBME)*, Feb 2011, pp. 134–137
21. Michigan State University. FOCUS. http://www.egr.msu.edu/~fultras-web/download.php
22. K. McLeish, D.L.G. Hill, D. Atkinson, J.M. Blackall, R. Razavi, A study of the motion and deformation of the heart due to respiration. IEEE Trans. Med. Imaging **21**(9), 1142–1150 (2002)

Chapter 5
System Architecture: Transponder

Keywords Ultrasound · Acoustic pressure · Wireless power transfer (WPT) ·
Wireless energy transfer · Wireless communication · Energy harvesting ·
CMOS · Battery · Attenuation · Operating frequency · Transponder · Sensor ·
Backscattering · Load modulation · Piezoelectric transducer · Rectifier · Voltage
regulator · On–off keying (OOK) · Amplitude-shift keying (ASK) · Variable gain
amplifier (VGA)

In the previous chapter the use control unit architecture was presented. Here,
the architecture of transponder is presented. Specifically, the circuit description
addresses choices in the wireless energy transmission and half-duplex communication. Figure 5.1 shows the block diagram of the system architecture, where on the
left side there is the control unit (CU) and on the right side the transponder (TR).
The CU is made up of the following blocks: the power management unit (PMU),
two FPGAs (master and slave configuration), 64 power amplifiers (PAs) and high-
voltage switches, 8 low-noise amplifiers (LNAs), and the demodulation circuit. On
the other side, the TR is made up of: the energy harvesting circuit (AC-to-DC and
DC-to-DC converter, battery charger, and battery), a μprocessor, a sensor, and the
modulation/demodulation circuit.

The control unit is used as mean to access the information stored into a transponder, which is implanted deeply in the human body. To retrieve this information, first,
the transponder needs to be energized and then, to be addressed [1]. The implanted
device is able to send to the control unit the stored data relative to the sensor
activity. Figure 5.2 shows a block diagram of the implanted medical device that
uses a μcontroller to synchronize the operations as the recharge of the μbattery and
the wireless communication. The recharge of the battery is enabled via the signal
En_EnHarv when it is set to ground (or default condition). Once the battery is fully
charged, the wireless communication can start and it is controlled via the signals
En_Comm_PMOS and *En_Comm_NMOS*. A parallel inductor L_P is used to cancel
out any imaginary part as the reactive component of the piezoelectric transducer
and the input capacitance of the entire system. Two zener diodes (D_Z) are used
as protection in order that the input voltage remains below 20 Vpp. An AC-to-DC

© Springer Nature Switzerland AG 2020   61
F. Mazzilli, C. Dehollain, *Ultrasound Energy and Data Transfer for Medical Implants*,
Analog Circuits and Signal Processing, https://doi.org/10.1007/978-3-030-49004-1_5

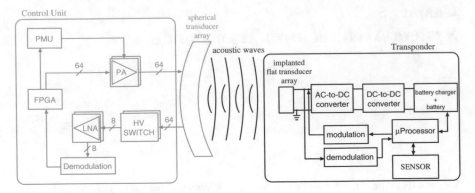

Fig. 5.1 System architecture block diagram

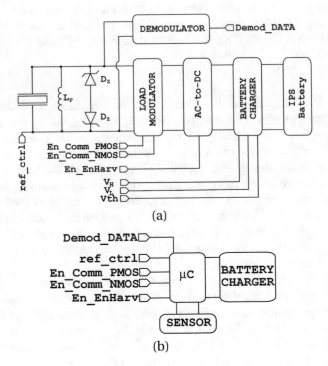

Fig. 5.2 Block diagram of the (**a**) transponder and (**b**) μcontroller connections

converter is employed to rectify the input sinusoidal signal so that the IPS battery can be recharged, as will be discussed in Sect. 5.2. The load modulator is connected directly to the piezo harvester, as will be discussed in Sect. 5.7.

Next section introduces an equivalent circuit of a piezoelectric transducer.

5.1   Equivalent Circuit of a Piezoelectric Transducer

Energy harvesting techniques exploit the use of piezoelectric materials [e.g. Lead-Zirconate-Titanate (PZT), polyvinylidene fluoride (PVDF), etc.] and their ability to transform mechanical stress (or ambient energy) into an electric potential. Conversely, an electrical charge can be converted into mechanical energy (e.g. microphones, accelerometers, ultrasonic transducers, etc.). A commonly used method for analysis of electro-mechanical systems is the equivalent circuit, which, when substituted for the system in any electric circuit, has the same effect as the system itself, at least in a limited frequency range.

5.1.1   Three-Port Network Model

For simplicity of reasoning an ultrasonic transducer (or resonator) is represented by a piezoelectric material of thickness d_0 and two metal electrodes created through deposition of metal film. In addition, to extend the bandwidth of the system, and therefore decreasing the quality factor, a backing material (or damping material) bonded to one side of the transducer is used, as illustrated in Fig. 5.3.

The piezoelectric transducer could be seen as a three-port network having two mechanical ports (or acoustic ports) and one electrical port, as shown in Fig. 5.4. One mechanical port is usually in contact with the medium (e.g. water, body) and the other mechanical port is usually coupled to the backing material.

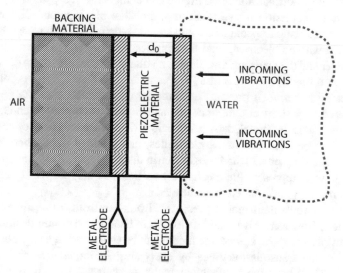

Fig. 5.3 Piezoelectric transducer loaded on its back surface

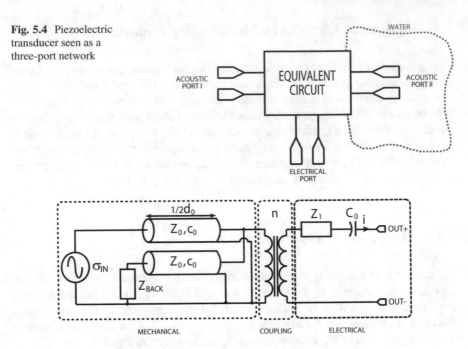

Fig. 5.4 Piezoelectric transducer seen as a three-port network

Fig. 5.5 KLM-based equivalent circuit of an ultrasonic transducer

The link from the incoming acoustic vibrations to the electrical port can be modeled with the help of transmission lines and a transformer used to couple the mechanical and electrical domains. Figure 5.5 depicts the KLM model [2], where the acoustic transmission lines are sized as half the piezo crystal thickness (d_0), having a characteristic impedance Z_0 and sound velocity c_0. Each transmission line is terminated with an electrical load that mimics the mechanical interface between the transducer and the environment. Hence, a voltage source represents the incoming vibrations as a function of the stress (σ_{IN}) and an impedance Z_{BACK} corresponds to the backing material (if present) or to the air. In the latter case, Z_{BACK} can be approximated as a short circuit since acoustic waves are highly attenuated in air. The electrical side is modeled by a capacitor C_0, which consists of a dielectric (or piezo) between the two metal electrodes, and a series impedance Z_1 due to the acoustoelectric feedback and the dielectric displacement. The transmission line analogy allows to represent the mechanical sections, so that the KLM network can be used to model a multi-layers transducer.

KLM-based equivalent model presents advantages from a transducer designer point of view. However, this model is difficult to handle in a circuit design software due to transmission lines. Hence, the model can be simplified to a two-port network by replacing the transmission lines by an equivalent impedance as it is shown in Fig. 5.6 [3]. Where the mechanical realm is replaced by L_M represents the mechanical mass, C_M the elasticity (or mechanical stiffness), and R_M takes into

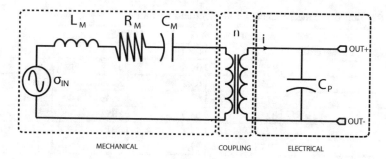

Fig. 5.6 Lumped-based equivalent circuit of an ultrasonic transducer

Fig. 5.7 Equivalent circuit of
a piezoelectric transducer

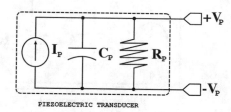

account the mechanical losses. A transformer (n) is again used to transform the mechanical stress to current (i). The electrical domain is represented by a parallel capacitor C_P which represents the plate capacitance of the piezoelectric material.

However, the equivalent model shown in Fig. 5.6 requires verilog behavioral modeling of the mechanical side.

5.1.2   Electrical Model

The circuit in Fig. 5.6 can be further simplified when the piezo is excited by sinusoidal vibrations (σ_{IN}) at frequency close or at resonance [4, 5] as is shown in Fig. 5.7. Here, the resistor R_P represents the internal resistance of the piezo which limits the amount of sinusoidal current (i_P) yields by the piezo and I_P the current amplitude.

The electrical impedance of the flat array is measured with a network analyzer (4294A) by coupling the array to water and connected to the network analyzer with a 1 m multicoaxial cable. The cable impedance was compensated to get the electrical impedance of each element. Table 5.1 shows the impedance measurement for the flat transducer array with and without the compensation of the coaxial cable. The average impedance of the array with the compensation of the coaxial cable is equal to $R_P = 3460\,\Omega$ and $X_P = (-j2550)\,\Omega$ and the coupling between the elements is $-54\,\text{dB}$ at 1 MHz (measurement performed by IMASONIC). The average impedance of the array without the compensation of the coaxial cable is equal to $R_P = 3552\,\Omega$ and $X_P = (-j614)\,\Omega$. To extract the maximum power

Table 5.1 Impedance measurement for the flat transducer array

# transducer element	Without coax cable		With coax cable	
	R_P (Ω)	C_P (pF)	R_P (Ω)	C_P (pF)
1	3575	64.6	4107	260.4
2	3366	65.2	3334	263.4
3	3125	66.3	3508	259.6
4	3280	63.9	3053	264.1
5	3727	59.1	3915	253.7
6	2773	57.4	3390	253.9

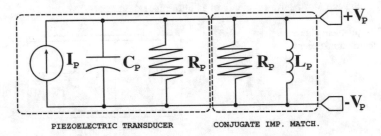

PIEZOELECTRIC TRANSDUCER   CONJUGATE IMP. MATCH.

Fig. 5.8 A conjugate impedance match for maximum power extraction

Table 5.2 Parallel inductor computed as $L_P = 1/(\omega^2 C_P)$ for $\omega = 1\,\text{MHz}$

# transducer element	Without coax cable		With coax cable	
	L_P (μH)	C_P (pF)	L_P (μH)	C_P (pF)
1	392	64.6	97.3	260.4
2	388	65.2	96.2	263.4
3	382	66.3	97.6	259.6
4	397	63.9	95.9	264.1
5	429	59.1	99.8	253.7
6	441	57.4	99.7	253.9

available out of the piezo, the interface (or AC-to-DC converter) circuit should present a conjugate impedance match respect to the piezo impedance as shown in Fig. 5.8.

The parallel inductor if such a conjugate impedance match is used, assumed a value equal to $L_P = 1/(\omega^2 C_P)$ (see Table 5.2), and the theoretical maximum power that can be extracted from the piezoelectric harvester can be expressed as

$$P_{max} = \frac{I_P^2 R_P}{8} = \frac{Q_P^2 V_P^2}{8 R_P} \tag{5.1}$$

where V_P is equal to $I_P/(\omega C_P)$ and the term $Q_P = \omega C_P R_P$ is the quality factor of the piezoelectric element.

5.1.3   Measurement and Simulation Results

The equivalent model shown in Fig. 5.8 is validated by measuring the voltage between the terminals $\pm V_P$ as a function of the load impedance. Figure 5.9 shows the circuit used to characterize the piezo output voltage characteristic. The third element of the flat array is considered ($R_P \approx 3510\ \Omega$, $C_P \approx 260\,\mathrm{pF}$), a coaxial cable of 1 m is connected between the piezo and the electrical circuit, which is made up of an off chip inductor $L_P = 100\,\mu\mathrm{H}$, a sense resistor $R_{SENSE} = (11.4)\ \Omega$, and the load resistor R_{LOAD}. The Agilent impedance analyzer 4285A is used to characterize R_{SENSE} and R_{LOAD} at 1 MHz. The voltages V_A and V_B are measured with the Agilent oscilloscope $DS06014A$, thus the sinusoidal current i_{LOAD} is indirectly measured to check that resonance occurs. The control unit board (Fig. 4.21) is used to send power towards the piezo harvester, which is located at an unknown position for the moment. Twenty-four power amplifiers are ON whose supply voltage is set to $V_{CC} = 2.2$ V while consuming a total power of 0.8 W.

Figure 5.10 shows the simulated and measured load current i_{LOAD} characteristic starting at the short-circuit current (I_{SC}) point. Hence, to check if the matched condition is achieved, i_{LOAD} is compared to $I_{SC}/2$ and the corresponding R_{LOAD} is compared to R_P. The simulated I_{SC} is equal to 6.2 mA at $R_{LOAD} = 0\ \Omega$, while the measured i_{LOAD} is equal to 5.9 mA at $R_{LOAD} = 2.73\ \Omega$. Therefore, the simulated $I_{SC}/2 = 3.1$ mA at $R_{LOAD} = 3510\ \Omega$ and the measured i_{LOAD} is equal 3.07 mA at $R_{LOAD} = 3440\ \Omega$. Hence, a good agreement between the measured and simulated results is shown, even if in presence of a small deviation.

Figure 5.11 shows the simulated and measured output voltage V_A characteristic, where the simulated open-circuit voltage V_{OC} is equal to 21.4 V at $R_{LOAD} = 10\,\mathrm{M}\Omega$, and the measured V_{OC} is equal to 21.4 V at $R_{LOAD} = 330\,\mathrm{k}\Omega$. To check if the matched condition is achieved, V_A is compared to $V_{OC}/2$ and the corresponding R_{LOAD} is compared to R_P. Therefore, the simulated $V_{OC}/2 = 10.7$ V at $R_{LOAD} = 3510\ \Omega$ and the measured V_A is equal 10.6 V at $R_{LOAD} = 3440\ \Omega$. Hence, a good agreement between the measured and simulated results is shown, even if in presence of a small deviation.

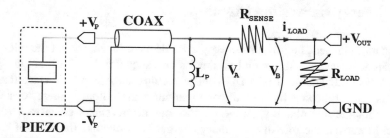

Fig. 5.9 Circuit used to validate the piezoelectric equivalent model

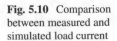

Fig. 5.10 Comparison between measured and simulated load current

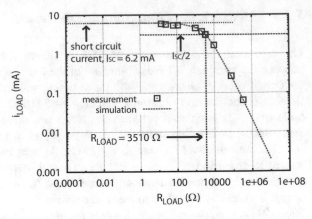

Fig. 5.11 Comparison between measured and simulated output voltage V_A

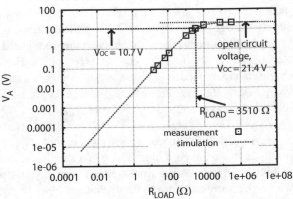

The equivalent model for the piezo harvester presented in Fig. 5.8 is thus validated. In the following section the energy harvesting principle along with battery charger is discussed.

5.2   Energy Harvesting Circuit

The control unit sends ultrasound energy towards the implant, so that the microbattery can be recharged. The main building blocks, shown in Fig. 5.12, are designed toward this end. The need of a DC voltage requires an AC-to-DC converter; this stage is often followed by a regulator to stabilize the output voltage to a value suitable for the microbattery, as the rectifier output presents a consistent voltage ripple. Moreover, the role of the battery charger is to monitor the state-of-charge of the microbattery in order to prevent that over charge from occurring. During the usual activities of the implant, such as wireless communication and acquiring signal

Fig. 5.12 Block diagram of
the energy harvesting system

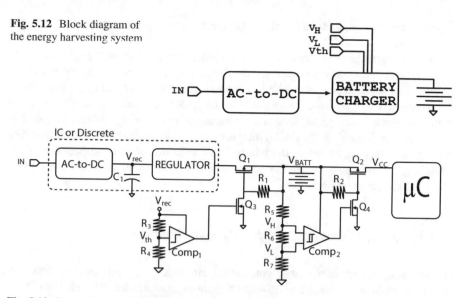

Fig. 5.13 Block diagram of the battery charger

from the sensor, the battery is discharging. This battery charger has to prevent the microbattery from discharging below a minimum threshold.

Figure 5.13 shows an insight of the battery charger, comparator $Comp_1$ checks that the output voltage of the regulator is equal to Vth (4.1 V) before connecting the battery to the regulator. While comparator $Comp_2$ checks that the battery voltage (V_{BATT}) stays between V_H (4.1 V) and V_L (2.5 V). If it happens that V_{BATT} falls out of this range, the microbattery can be damaged and thus it is immediately disconnected from the μcontroller. $Comp_2$ is powered directly from the battery since it needed to stays always ON: it consumes 2 μA.

The AC-to-DC converter and the voltage regulator are reviewed next.

5.3   Design Methodology and Comparison of Rectifiers

In this subsection, it is discussed how an integrated rectifier for UHF frequencies can be designed to match the transducer impedance without adding external components. Nevertheless the transducer needs to have an imaginary part that presents an inductive behavior as the input impedance of the rectifier acts as a capacitor. In case of an electroacoustic transducer that operates at low frequency (e.g. 1 MHz) and shows a capacitive behavior, the matching network has to be off chip. Figure 5.2a depicts a parallel inductor L_P used to create a parallel RLC network. Hence, a design of a matching network is not considered in this work since the flat transducer used for the transponder is realized by six elements (see Fig. 3.14), and requires six off-chip inductors. However, the constraints to design a discrete rectifier are relaxed for the impedance matching network since it requires at least two elements [6].

To maximize the available power at the energy storage element tag it is necessary to minimize the power loss across the rectifier which is achieved by matching the rectifier input impedance with the antenna [7]. Moreover, using an impedance matching network, tag read range increases by boosting the available voltage to the rectifier even in cases where the available input power is low [8]. Hence, the rectifier power conversion efficiency (PCE) increases as the available voltage to the rectifier increases. However, to select the nature and the values of the components of the matching network an accurate derivation of the rectifier input impedance has to be done through circuit simulation.

5.3.1   Design Methodology for Passive Rectifier

To increase the rectifier power conversion efficiency the rectifier impedance is to be matched with the antenna for maximum power transfer. The rectifier input admittance is modeled as a parallel combination of a capacitor and a resistor mathematically represented as $Y_{rec} = G_{rec} + j*Y_{Crec}$, where G_{rec} and Y_{Crec} represent the nonlinear input conductance and susceptance, respectively [1]. To match the rectifier input impedance with the antenna (in this case a $50\,\Omega$ characteristic impedance is considered) the imaginary part is compensated with a parallel inductor (L_p) which is represented by an equivalent series inductance (L_s) as shown in (Fig. 5.14). A series inductor topology is used to compensate the rectifier capacitive part as it boosts the voltage across the rectifier by a factor "$Q = \omega C_{rec}/G_{rec}$" (which is the resonant structure quality factor) mathematically represented in (5.2), where V_{AV} corresponds to the input RF voltage as shown in Fig. 5.14 thereby increasing the voltage conversion efficiency. Once the capacitive part is compensated by the inductor the rectifier nonlinear resistance is matched to $50\,\Omega$ by having proper transistor dimensions (i.e. W/L ratio).

$$|V_{IN}| = \frac{|V_{AV}|}{2}\sqrt{1 + Q^2} \tag{5.2}$$

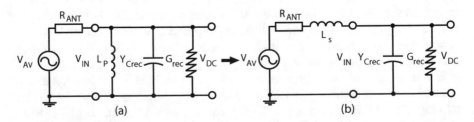

Fig. 5.14 Simplified schematic of the tag including the antenna, the rectifier equivalent circuit, and matching network solutions through an inductor: (**a**) shunt, (**b**) series

Fig. 5.15 NMOS bridge
rectifier circuit

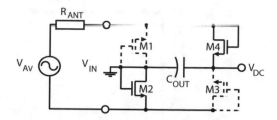

In the case of MOSFET based diode structures, to obtain the series inductance value (L_s) it is necessary to compute the rectifier input capacitance which in turn depends on the transistor dimensions for a fixed gate oxide capacitance. The analytical derivation for the rectifier input nonlinear resistance is determined based on the I–V relationship of the MOS device which becomes complicated due to the short channel effects and the quadratic dependency between the current and the voltage. Therefore the aide of large-signal analysis in the simulator is used to determine the rectifier input impedance.

The design procedure for the bridge rectifier shown in (Fig. 5.15) is described. During the positive cycle of the input RF signal V_{AV} the transistors M4 and M2 are switched on and during the other cycle transistors M3 and M1 are switched on charging the capacitor C_{OUT} in a single direction thereby rectifying the input RF signal.

The input nonlinear resistance determined for a given input power (worst case scenario) using periodic steady state (PSS) analysis in Cadence simulator for various values of W/L is shown in the left-hand y-axis of Fig. 5.16 and the corresponding value of the impedance of the capacitor is shown in the right-hand y-axis of Fig. 5.16. From Fig. 5.16 the input capacitance to achieve 50 Ω is calculated and then the capacitance is compensated by a corresponding series inductance value for the given operating frequency to boost up the voltage across rectifier.

Figure 5.17 compares the rectifier input resistance measured using a VNA (Agilent HP8719D) with the results obtained using Cadence Virtuoso simulator. From Fig. 5.17 it can be seen that the measured value is in good agreement with the simulated value thereby validating the design procedure for the rectifiers.

Figure 5.18 shows the measured voltage efficiency as a function of the input available voltage V_{AV}. An off-chip series inductor $L_s = 12$ nH was used to boost the input voltage of the rectifier at input RF signal frequency of 900 MHz. From Fig. 5.18 it can be seen that the rectifier voltage conversion efficiency matched to the antenna is higher in comparison to that without the series inductor, which is as expected. For the calculations, the rectifier voltage conversion efficiency (VCE) is defined as:

$$\eta = \frac{V_{DC}}{V_{AV,p}} \tag{5.3}$$

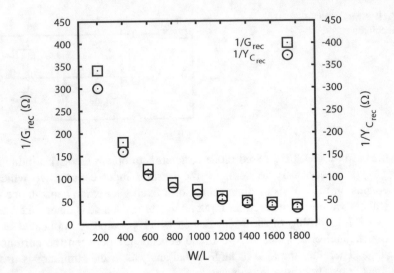

Fig. 5.16 Simulated nonlinear resistance and impedance of the capacitor of the NMOS bridge rectifier

Fig. 5.17 Simulated and measured real part of the rectifier input impedance with transistor size ratio $W/L = 500$

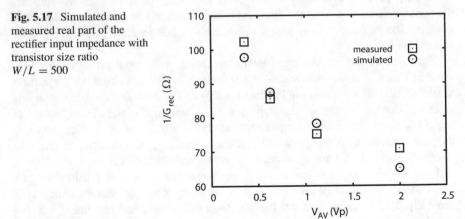

5.3.2 Comparison of Rectifiers

Optimization of the single-stage rectifier is crucial to improve the voltage conversion efficiency (VCE) that can be increased by cascading several single-stage rectifiers. Four rectifier configurations shown in (Fig. 5.19) are compared using VCE as the Figure-of-Merit (FoM). The rectifier configurations are implemented in CMOS $0.18\,\mu m$ technology using zero-Vth transistors (Vth $= 2\,mV$) for NMOS, and low-Vth transistors (Vth $= 320\,mV$) for PMOS. The use of zero-Vth and low-Vth decreases the drop across the transistors and thus increases the voltage

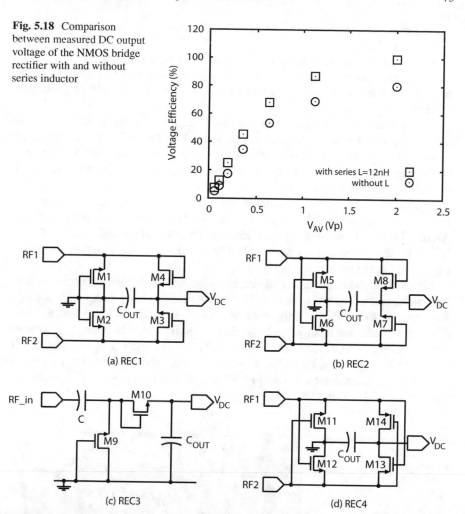

Fig. 5.18 Comparison between measured DC output voltage of the NMOS bridge rectifier with and without series inductor

Fig. 5.19 Rectifier topologies: (**a**) NMOS differential-drive bridge rectifier, (**b**) NMOS differential-drive gate cross-connected bridge rectifier, (**c**) NMOS doubler, (**d**) NMOS-PMOS differential-drive gate cross-connected bridge rectifier

conversion efficiency. A short description for each rectifier topology along with the experimental results is described below.

Figure 5.19a shows the NMOS differential-drive bridge rectifier normally found in RFID applications where the reader and the tag are in close proximity. For proper rectification using such a topology the input voltage across the rectifiers should be at least twice the threshold voltage, hence using zero-Vth transistors enables such an architecture to operate even at low input power levels.

Figure 5.19b is a modified version of the common bridge rectifier, where the NMOS transistors whose gates were grounded, in Fig. 5.19a, are now cross-coupled to the inputs thereby decreasing the required input voltage to turn on the transistors and hence an increase in the read range of the RFID tag.

Figure 5.19c is a well-known doubler structure where the output V_{DC} ideally is equal to twice the peak input RF voltage. Many such stages can be used in cascade to increase the output voltage hence increasing the overall voltage conversion efficiency.

Figure 5.19d has its input stage similar to Fig. 5.19b. However, in this case the PMOS transistors are also cross-coupled and hence all four transistors act as switches. A major disadvantage of such a topology is the need for a current controlling circuitry as the current direction reverses when the rectified voltage is higher than the input RF voltage.

Figure 5.20 shows the photomicrograph of the four rectifiers, fabricated in 0.18 μm CMOS process. The chip was glued on a PCB used for testing the rectifiers. All transistors have width and length 250 μm/0.5 μm, and output capacitor, C_{OUT}, is 5 pF for the bridge converters and 10 pF for the doubler. The input capacitance in Fig. 5.19c has the same value as C_{OUT}.

Figure 5.21 shows the unloaded voltage efficiency of the fabricated single-stage RF-to-DC converters as a function of the available voltage along with a signal frequency of 900 MHz. The VCE increases more than 100% for Fig. 5.19b, d, which is the result due to the input voltage boosting due to the series inductor (L_s), whereas for Fig. 5.19a, c, VCE increases toward 90%. Three regions can be identified in Fig. 5.21: $V_{AV} < 0.5$ V, Fig. 5.19d works efficiently; then, for $0.5\,\text{V} \leq V_{AV} \leq 1\,\text{V}$ there is a transition zone between Fig. 5.19d and b for $V_{AV} > 1$ V, the output DC voltage of Fig. 5.19b increases above V_{AV}.

Fig. 5.20 Photomicrograph of fabricated chip in 0.18 μm CMOS process

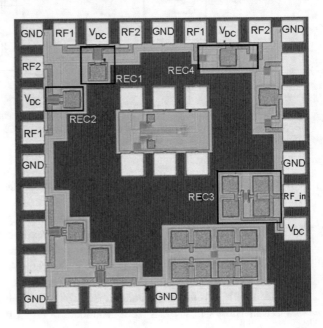

Fig. 5.21 Measured unloaded voltage efficiency of the RF-to-DC converters fabricated in 0.18 μm CMOS process

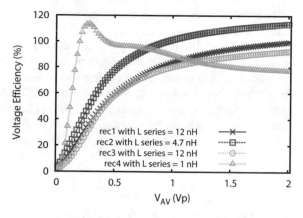

Fig. 5.22 Measured loaded voltage efficiency of RF-to-DC converters at input available voltage 110 mV

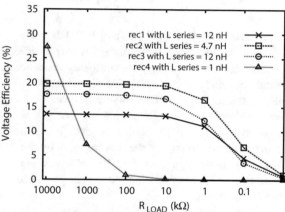

Figure 5.22 shows the voltage efficiency of the RF-to-DC converters under different loads with V_{AV} =110 mV at 900 MHz. Even if Fig. 5.19d for low input available voltage shows the best VCE, in loaded condition its efficiency decreases drastically. Whereas for the other topologies the VCE is constant until 10 kΩ and Fig. 5.19b clearly has the best loaded voltage efficiency.

5.3.3 Summary

A design and comparison methodology for different RF-to-DC converters in UHF-band for RFIDs has been proposed. The input impedance obtained using large-signal simulation is compared with the measured results. Four different RF-to-DC converters have been designed and fabricated in 0.18 μm CMOS process. FOM as voltage efficiency has been given to compare the performances of the rectifiers. This analysis allows the designer to select the proper rectifier according to its operating condition defined by the available input voltage. However, as an HV CMOS technology has been chosen to realize the rectifier for the transponder, the rectifier in Fig. 5.19c should be discarded since it is required to overcome the Vth of two transistors before rectification occurs. In addition, the rectifier in Fig. 5.19b may be used since it is required to overcome the drain-to-source voltage, which is lower than the threshold voltage of the transistor. In XFAB XH018 CMOS technology zero-Vth transistors are not available so it could be better to employ a rectifier that uses a single switch. As the six elements of the flat transducer array lie on the same substrate, a single-ended rectifier is designed for each element as it will be explained later in this book.

5.4   Design of an Active Rectifier

In far-field battery-less (or passive) RFID systems, a control unit (or base station) sends energy towards a transponder (or tag) in order to be able to read (or retrieve) the data stored in a memory. A multi-stage rectifier [or voltage multiplier (VM)] is usually designed to convert the RF signal to DC voltages when the received power falls in the μW range. The voltage doubler (see Fig. 5.19c) is the basic building block to implement the Dickson's rectifier [9], which is adapted to suit the need of passive RFID systems [10]. Moreover, a threshold compensation technique can be implemented to reduce the loss imposes by the turn-on voltage of the transistor [11]. As explained in Sect. 5.3, the antenna could be co-designed along with the rectifier to boost the available input voltage and to avoid the use of additional off-chip components.

In case of emerging applications like wireless micro-sensor networks [12], where a base station is impractical, for reason such as been too far to provide enough power to energize a sensor node, ambient vibrations (or ambient noise) can be a good alternative since this available energy is free. For instance, a piezoelectric energy harvester subjected to ambient vibrations (or mechanical energy) can yield an output power of 10–100 μW [13, 14]. Hence, considerable work has been done on interface circuits for piezoelectric harvesters to maximize the energy conversion (or extraction) [15, 16]. However, CMOS voltage doublers (see Fig. 5.19c) and full-bridge rectifiers (see Fig. 5.19a) can extract only 12.5% of the actual power available from the piezoelectric harvester. Furthermore, it has been shown that a bias-flip rectifier can improve the power extraction up to 81% using conjugate impedance matching [17]. Figure 5.23 shows the bias-flip rectifier, where a simple switch m_1 in series with an inductor L_{BF} is connected across a full-bridge rectifier (D_1–D_4). The inductor helps to reverse (or flip) the voltage across the capacitor when connected to it via m_1, so that the current i_P has to complete the charge (or discharge) of C_P up to $\pm(V_{rec} + 2V_D)$ [18].

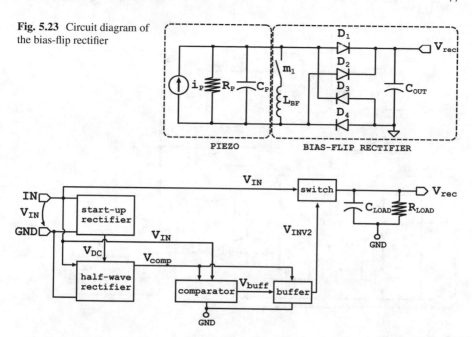

Fig. 5.23 Circuit diagram of the bias-flip rectifier

Fig. 5.24 Block diagram of the single-stage active rectifier

The bias-flip rectifier extracts energy from both positive and negative cycles of the piezo harvester to increase the power efficiency. In the same manner, differential synchronous (or active) rectifiers have been developed for inductively powered devices. Therefore, a lot of work has been done to design high efficient power converter for high- and low-power applications [19–26]. Although differential mode rectifiers show high power conversion efficiency under light- or heavy-load conditions, a single-ended AC-to-DC converter can offer the same performances [27, 28]. Therefore, a single-ended synchronous (or active) rectifier topology is envisaged since the six crystals (see Fig. 3.14b) lie on the same substrate.

5.4.1 A Novel Synchronous Rectifier

Figure 5.24 shows the block diagram of the single-stage active rectifier made up of a start-up rectifier, a half-wave rectifier, a comparator, a buffer, and a power switch. The start-up rectifier yields an output V_{DC} that powers a half-wave rectifier, whose output V_{comp} is compared with the input signal V_{IN}. The output of the comparator V_{buff} is amplified and squared via the buffer, whose output V_{INV2} activates the switch, which generates the rectified voltage V_{rec}.

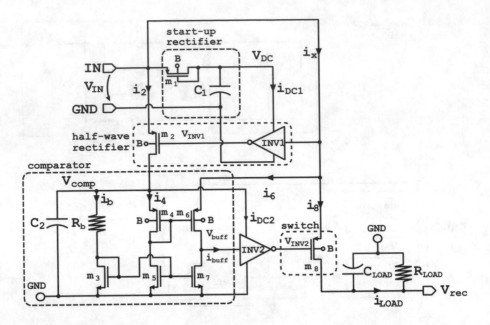

Fig. 5.25 Circuit diagram of the single-stage active rectifier

Figure 5.25 shows the circuit diagram of the single-stage active rectifier. The start-up rectifier is made up of a diode-connected transistor m_1 and a capacitor C_1 which yields an output voltage V_{DC} and a current i_{DC1} to power the inverter $INV1$. The voltage conversion efficiency (VCE) for the start-up rectifier can be expressed as

$$VCE_{st\_up} = \frac{V_{DC,p}}{V_{IN,p}} = \frac{V_{IN,p} - |V_{t,1}|}{V_{IN,p}} \tag{5.4}$$

where $V_{IN,p}$ represents the input peak voltage, $V_{DC,p}$ the output peak voltage, and $V_{t,1}$ the threshold voltage of the transistor m_1. As shown by (5.4), the efficiency can be improved by choosing a transistor with low V_t.

High-voltage transistors are employed to build the interface circuit for the piezo harvester, so the VCE_{st_up} results to be low. However, this topology can be useful to provide energy to the active half-wave rectifier [22], which voltage conversion efficiency during the ON state of the transistor m_2 can be expressed as

$$VCE_{h\_w} = \frac{V_{comp,p}}{V_{IN,p}} = \frac{V_{IN,p} - |V_{DS,2}|}{V_{IN,p}} \tag{5.5}$$

where $V_{DS,2}$ is the drain-to-source voltage drop across the transistor m_2 and V_{comp} is the rectified output peak voltage which is used to power the comparator and a buffer $INV2$. Thus the voltage conversion efficiency VCE_{h_w} results to be higher than VCE_{sl_up} as well as $V_{comp,p}$ is higher than $V_{DC,p}$ since the drain-to-source voltage V_{DS} may be lower than the threshold voltage V_t. V_{DS} for the transistor m_2 during the ON state can be approximated as

$$|V_{DS,2}| \approx \frac{i_{2,max}}{k_{p,p}\left(\frac{W}{L}\right)_2 \left(|V_{INV1} - V_{IN,p}| - |V_t|\right)} \tag{5.6}$$

where $i_{2,max}$ represents the peak current through m_2 which causes the maximum voltage drop across the drain and source terminals, $(W/L)_2$ and $k_{p,p}$ the width, the length, and the gain factor of the transistor and V_{INV1} the output voltage of the buffer $INV1$ which is assumed equal to $0\,V$ when m_2 is conducting. The ratio $|V_{DS,2}|$ over $i_{2,max}$ describe the on-resistance as follows

$$R_{ON} \approx \frac{|V_{DS,2}|}{i_{2,max}} = \left[k_{p,p}\left(\frac{W}{L}\right)_2 \left(|V_{INV1} - V_{IN,p}| - |V_t|\right)\right]^{-1} \tag{5.7}$$

By combining (5.5) and (5.7), V_{comp} can be approximated as

$$V_{comp} \approx V_{IN,p} - R_{ON}i_2 \tag{5.8}$$

to increase V_{comp} the second term in (5.8) has to be small. The maximum value of V_{comp} ($V_{comp,max}$) is determined when i_2 is maximum, while the minimum value ($V_{comp,min}$) is determined during the off-state of the transistor m_2 by discharging the capacitor C_2. Therefore, the voltage ripple is a design parameter and can be expressed as

$$\Delta V_{comp} = V_{comp,max} - V_{comp,min} \tag{5.9}$$

The capacitor C_2 should ensure a voltage ripple lower than ΔV_{comp} and its value is found as [29]

$$C_2 \geq \frac{i_{2,max}T}{\Delta V_{comp}}\left(\frac{1}{k} - \frac{\delta}{2\pi}\right) \tag{5.10}$$

where T represents the period (1 µs) of the input sinusoidal current i_p (Fig. 5.7), k the number of rectifying alternation (1 in this case), and δ the conducting angle. To simplify the design procedure (5.10) is approximated as

$$C_2 \geq \frac{i_{2,max}T}{\Delta V_{comp}} \tag{5.11}$$

To assure a DC bias point for the comparator, $V_{comp,min}$ is selected as follows

$$V_{comp,min} = R_b i_b + V_{GS,3} = R_b i_b + \sqrt{\frac{2i_b}{k_{p,n}(W/L)_3}} + V_{t,3} \tag{5.12}$$

where i_b represents the biasing current of the comparator, R_b the biasing resistor, $V_{GS,3}$ the gate-to-source voltage of m_3, $(W/L)_3$, $V_{t,3}$ and $k_{p,n}$ the width, the length, the threshold voltage, and the gain factor of the transistor. To assure that m_4 and m_5 turn on, $V_{comp,min}$ can be rewritten as

$$V_{comp,min} = |V_{t,4}| + V_{t,5} \tag{5.13}$$

where $V_{t,4} = V_{t,3}$ and $V_{t,5}$ are the threshold voltages of m_4 and m_5, respectively. By combining (5.12) and (5.13), the lower bound for R_b is expressed as

$$R_b \geq \frac{|V_{t,4}|}{i_b} - \sqrt{\frac{2}{k_{p,n}(W/L)_3 i_b}} \tag{5.14}$$

Hence, i_b is mirrored into m_5 and m_7 via m_3; thus if V_{IN} is higher than V_{comp}, then $i_6 \approx i_{buff}$ charges the input capacitor of $INV2$ whose output V_{INV2} is set to zero. On the contrary, if V_{IN} is lower than V_{comp}, V_{INV2} is set to one [30].

High-voltage transistors are used to implement the single-stage active rectifier, which have a minimum channel length of 2.9 µm, threshold voltages of $V_{t,n} = 1.55$ V and $|V_{t,p}| = 1.42$ V for a NMOS and a PMOS transistor, respectively. Therefore, using (5.13), $V_{comp,min}$ is equal to 3 V. The biasing currents are chosen as shown in Table 5.3(a), a maximum peak current for i_2 is selected equal to 2 mA since powers the buffer $INV2$; while the biasing current i_b is determined at $V_{comp,min} = 3$ V. Table 5.3(b) shows the width and length for each transistor.

The passive components C_1, C_2, R_b, and C_{LOAD} are chosen equal to 2.7 pF, 90 pF, 50 kΩ, and 100 nF, respectively. Figure 5.25 shows a floating bulk B for all the PMOS transistors, thus the well-known adaptive-body-bias (ABB) technique is

Table 5.3 Design variables

(a) Biasing currents		(b) Transistors size			
Current	mA	Transistor	W (µm)	L (µm)	W/L
$i_{DC1,max}$	0.06	m_1	50	2.9	17
$i_{2,max}$	2	m_2	100	2.9	34.5
i_b	0.01	m_3	6	20	0.3
i_4	0.01	m_4	80	2.9	27.6
$i_{8,max}$	20	m_5	6	20	0.3
$i_{LOAD,max}$	20	m_6	80	2.9	27.6
		m_7	6	20	0.3
		m_8	1800	2.9	620

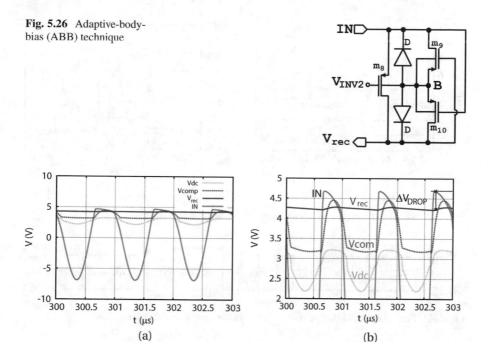

Fig. 5.26 Adaptive-body-bias (ABB) technique

Fig. 5.27 (a) Simulated voltage waveforms of the single-stage active rectifier. (b) Voltage drop from input (V_{IN}) to output (V_{rec}) equal to $\Delta V_{DROP} = 460\,\text{mV}$

implemented as sketched in Fig. 5.26 [31, 32] to prevent that the body diodes turn on D. The bulk is dynamically connected to the maximum voltage yields either at the terminal *IN* or V_{rec}.

5.4.2 Simulation Results

The circuit in Fig. 5.25 is simulated along with the piezo harvester equivalent model (see Fig. 5.7) that provides a peak current I_P equal to 5 mA. The lumped parameters R_P and C_P are measured by connecting the flat transducer array through a coaxial cable to the precision impedance analyzer 4294A from Agilent Technologies. At low frequency the impedance matching network can be avoided if the length of a cable (or wire connections) is less $\lambda/20$ [33]. Hence, the parallel capacitor C_P and parallel resistance R_P are equal to 260 pF and 4130 kΩ, including the coax cable impedance. A parallel inductor of 100 μH is used to cancel out the effect of the capacitor C_P at 1 MHz.

Figure 5.27a shows voltage waveforms of the single-stage active rectifier and Fig. 5.27b highlights the voltage drop from V_{IN} to V_{rec} when i_8 is maximum, thus $\Delta V_{DROP} = V_{IN} - V_{rec}$ is equal to 460 mV, one third of the threshold voltage ($|V_{t,p}| = 1.42$).

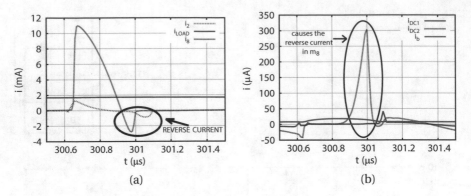

Fig. 5.28 Simulated current waveforms of the single-stage active rectifier

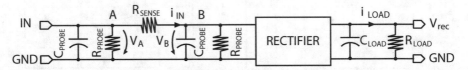

Fig. 5.29 Simple model of the rectifier to compute the power conversion efficiency

Figure 5.28 shows current waveforms when the rectifier is loaded with $2.5\,\mathrm{k\Omega}$ resistor. Reverse current for i_8 curve is present for a period of time about $100\,\mathrm{ns}$, due to the switching activity of the buffer *INV2* shown by the peak in i_{DC2}. Hence, an average current of $1.7\,\mathrm{mA}$ is delivered to R_{LOAD} at $4.3\,\mathrm{V}$ (V_{rec}).

Figure 5.29 shows the rectifier as two-port network.

Thus the voltage conversion efficiency (VCE) can be expressed as

$$VCE_{sync} = \frac{V_{rec}}{V_{B,p}} = \frac{V_{B,p} - \Delta V_{DROP}}{V_{B,p}} \qquad (5.15)$$

where V_{rec} represents the output voltage, $V_{B,p}$ the input peak voltage, and ΔV_{DROP} the voltage drop from the input node B to the output node V_{rec}. In addition, the power conversion efficiency (PCE) is seen as the ratio between the power delivered to the load and the input power which is expressed as

$$\overline{P_{IN}} = \frac{1}{T} \int_0^T V_B(t) \times \frac{(V_A(t) - V_B(t))}{R_{SENSE}} dt \qquad (5.16)$$

where T is the integration time, the difference $(V_A - V_B)$ is the voltage drop across the series resistor R_{SENSE} used to compute the input current to the rectifier i_{IN}. To measure the input current i_{IN} by indirect measurement thus sensing V_A and V_B is not recommended since oscilloscope probes introduce phase shifting between signals. A different approach need to be envisaged to determine the input power as phase shifting is a source of error.

5.4.3 *Experimental Results*

To check the performance of the rectifier (e.g. power conversion efficiency) along with simulation results, the piezo harvester is characterized as mentioned in Sect. 5.1. Thus the open-circuit voltage and the short-circuit current of the piezo are measured in order to set the available peak current I_P in Cadence environment. Figure 5.30 sketches the set-up used to characterize the crystal. The sense resistor R_{SENSE} is set to $10\,\Omega$ and the third element of the flat array is chosen to test the rectifier. The control unit board (Fig. 4.21) is used to send power towards the crystal, which is located at an unknown position for the moment. Twenty-four power amplifiers are ON whose supply voltage is set to $V_{CC} = 3.5\,V$ while consuming a total power of 3 W.

Figure 5.31 shows the simulated and measured current i_{IN} versus the load resistor R_P, whose value is swept from $50\,\Omega$ to $330\,k\Omega$ during measurements and from $1\,m\Omega$ to $10\,M\Omega$ during simulations. The short-circuit current (I_{SC}) is measured at $R_P = 50\,\Omega$, since a phase shift between V_A and V_B is present for lower value of R_P. To perform simulations in order to validate the model of the piezo harvester, the peak available current I_P is set equal to I_{SC}.

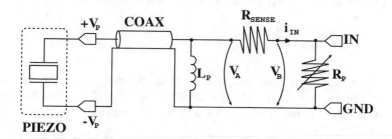

Fig. 5.30 Circuit used to check the available peak current for the rectifier

Fig. 5.31 The available peak current for the rectifier is 5 mA, if the rectifier is matched to the piezo

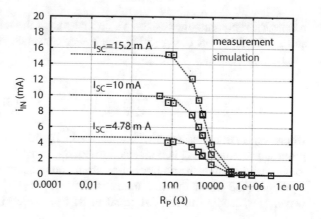

Fig. 5.32 The available peak
voltage for the rectifier, if the
rectifier is matched to the
piezo

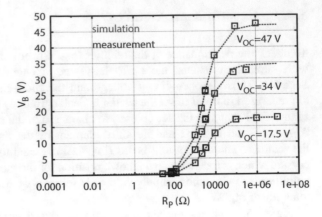

Figure 5.32 shows the simulated and measured voltage V_B versus the load resistor
R_P, whose value is swept from $50\,\Omega$ to $330\,\text{k}\Omega$ during measurements and from
$1\,\text{m}\Omega$ to $10\,\text{M}\Omega$ during simulations. The open-circuit voltage V_{OC} is measured at
$R_P = 330\,\text{k}\Omega$. Thus, the piezo is modeled as a current source ($i_P = I_P sin(\omega t)$) in
parallel with a capacitor and a resistor.

The available input power to the resistor R_P can be written as

$$P_{IN} = \frac{1}{2} \times i_{IN,p}^2 \times R_P = \frac{1}{2} \times \left(\frac{I_P}{2}\right)^2 \times R_P \tag{5.17}$$

where I_P is the piezo peak available current. Assuming that the rectifier is
conducting during only the positive phase of the input signal V_B, thus the average
input power can be found by solving the following integral as

$$\overline{P_{IN}} = \frac{2}{\pi} \int_0^{\pi/2} \frac{(I_P sin(x))^2 \, R_P}{8} dx = \frac{I_P^2}{16} \times R_P \tag{5.18}$$

where R_P represents the resistor value that match the resistive losses of the piezo
and x is equal to ωt. Using (5.18), the power conversion efficiency for the rectifier
can be expressed as

$$PCE = \eta_{rec} = \frac{P_{OUT}}{P_{IN}} = \left(\frac{V_{rec}^2}{R_{LOAD}}\right) \times \left(\frac{1}{P_{IN}}\right) \tag{5.19}$$

where R_{LOAD} represents the output load. To boost the output voltage of the
proposed rectifier, V_{rec} is connected to V_{comp}. In addition, the capacitor C_2 is no
longer required thus the number of external components is reduced.

Figure 5.33 shows the measurement of the input (V_B) and output (V_{rec}) voltage
waveforms, leading to a VCE equal to 81.5%, which is computed using (5.15).

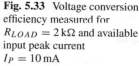

Fig. 5.33 Voltage conversion efficiency measured for $R_{LOAD} = 2\,k\Omega$ and available input peak current $I_P = 10\,mA$

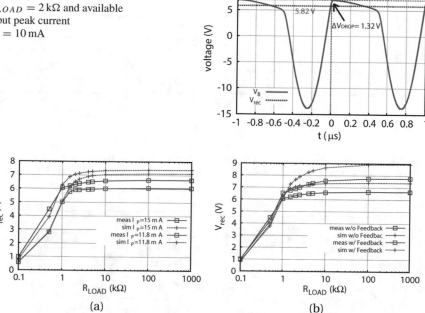

Fig. 5.34 Comparison between simulated and measured: (**a**) rectifier output voltage at different available peak current I_P and (**b**) with (w/) and without (w/o) the connection between V_{rec} and V_{comp} at I_P equal to 15 mA

However, the voltage drop across the rectifier is increased by 800 mV compared to simulation result.

The rectifier output voltage V_{rec}, output power P_{OUT}, and average input power $\overline{P_{IN}}$ are measured at different load (R_{LOAD}) values in order to derive the PCE as expressed in (5.19). The results are compared with simulated performances at the operating frequency of 1 MHz and available input peak current I_P.

Figure 5.34 shows the output voltage V_{rec} versus the load resistor R_{LOAD} swept from 100 Ω to 1 MΩ. The measured output voltage is lower than the simulated output waveform as mentioned previously an additional voltage drop is present across the rectifier. Figure 5.34b compares the output voltage of the rectifier with (w/) and without (w/o) the feedback from V_{rec} to V_{comp}. An improvement of 1 V is achieved by feeding the output back to V_{comp} which leads to more available current for the load.

Figure 5.35 shows the simulated and calculated average input power to the rectifier at different available input peak current I_P. The simulated average input power considers the current that flows through the rectifier while the calculated input power is $I_P^2 R_P / 16$.

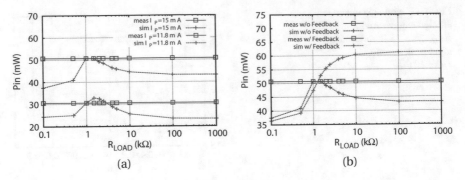

Fig. 5.35 Simulated and calculated average input power: (**a**) at different available input peak current I_P and (**b**) with (w/) and without (w/o) the connection between V_{rec} and V_{comp} at I_P equal to 15 mA

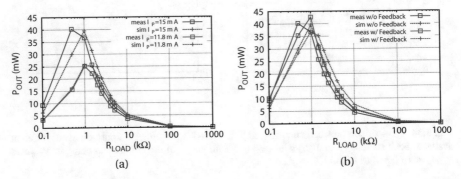

Fig. 5.36 Comparison between simulated and measured output power: (**a**) at different available input peak current I_P and (**b**) with (w/) and without (w/o) the connection between V_{rec} and V_{comp} at I_P equal to 15 mA

Figure 5.36 shows the delivered output power in mW versus the load resistor. Figure 5.36a shows the output power at 15 and 11.8 mA available input peak current (I_P). Figure 5.36a compares the output power in presence of voltage boosting thanks to the feedback connection from V_{rec} to V_{comp}. A good agreement is found between the simulated and measured characteristics.

Figure 5.37a plots the power conversion efficiency at I_P equal to 15 and 11.8 mA thus the measured PCE at $R_{LOAD} = 1\,k\Omega$ results equal to 72.38% and 82.45%, respectively. While the simulated PCE is 77.87% for $I_P = 15\,mA$ and 83.81% for $I_P = 11.8\,mA$. Figure 5.37b compares the power efficiency with and without the feedback connection for $I_P = 15\,mA$. With the feedback connection, PCE is equal to 84.19% from measurement and 77.77% from simulation.

Table 5.4 benchmarks recently reported active rectifiers in standard CMOS technology and a high-voltage passive rectifier used in various power management blocks along with the proposed synchronous rectifier. The active rectifiers presented in [20, 21, 27] and [23, 25, 26] make use of fast comparator to directly control the

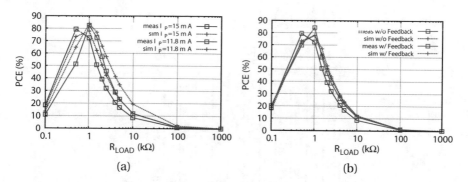

Fig. 5.37 Comparison between simulated and measured: (**a**) rectifier average input power and (**b**) power conversion efficiency (PCE). The available input peak current I_P is equal to $10\,\text{mA}$

gate of PMOS transistors to reduce reverse current leakage from the output load to the input, thus the same technique is employed in this work. The IC active area depends on the minimum channel length and on the width to length ratio. The power conversion efficiency (PCE) is above 80% as for the other topologies, while the voltage conversion efficiency (VCE) can be improved by increasing the W/L ratio. The PCE expressed in (5.19) can be rewritten as

$$PCE = \eta_{rec} = \frac{P_{OUT}}{P_{IN}} = \frac{P_{OUT}}{P_{OUT} + P_{LOSS}} = \frac{1}{1 + \frac{P_{LOSS}}{P_{OUT}}} \qquad (5.20)$$

where P_{LOSS} are the internal losses of the rectifier (e.g. comparator power consumption, switching loss). Hence, PCE can be improved by keeping the ratio P_{LOSS}/P_{OUT} as small as possible. In [20] $P_{LOSS} = 6.25\,\text{mW}$ and using (5.20), the theoretical PCE is equal to 87.8% which is close to the measured value 89%. Applying (5.20) in reference [27], the theoretical PCE is equal to 89.8% very close to the measured value 85%. Therefore, the total power loss for the single-stage synchronous rectifier, presented in this chapter, is derived using (5.20), thus P_{LOSS} is equal to 5.321 mW.

5.4.4 Summary

Passive rectifiers offer a good trade-off between delivered power and output voltage in low-power systems. Whereas the demand of power increases, an active rectifier design should be envisaged. Hence, active solutions can achieve high voltage and power conversion efficiency for remotely powered RF-, inductive-, and piezo-based link.

Table 5.4 Rectifier benchmarking

Publication	2004 [19]	2006 [21]	2006 [20]	2006 [27]	2008 [23]	2009 [24]	2011 [26]	2011 [25]	This work
CMOS Technology (μm)	1.5	0.35	0.5	0.25	0.5	0.35	0.5	0.8 HV	0.18 HV
Structure	Active	Active	Active	Active	Active	Active	Active	Passive	Active
$V_{IN\text{-}peak}$ (V)	N/A	3.5	5	N/A	5	2.4	3.8	15.68	6.54
V_{rec} (V)	<5	3.22	4.75	1.4	4.36	2.08	3.12	12	5.15
VCE (%)	N/A	92	95	N/A	87.2	87	82.1	76.5	78.74
R_{LOAD} (kΩ)	1	1.8	0.5	80	1	0.1	0.5	12	1
P_{OUT} (mW)	25	5.7	45	0.025	19	43	19.5	12	25
C_{LOAD} (μF)	0.01	0.0002	0.1	10	1	1	10	1	0.1
f (MHz)	1	13.56	0.5	<0.5	0.5	0.2	13.56	13.56	1
Area (mm^2)	0.48	0.0055	N/A	0.015	0.405	0.4	0.18	N/A	0.114
W/L (mm/μm)	4.8/1.6	0.24/0.35	N/A	0.6/0.6	2.5/0.6	N/A	2.1/0.6	N/A	1.8/2.9
PCE_{SIM} (%)	N/A	87	N/A	N/A	90.4	87	87	92.3	83.81
PCE_{MEAS} (%)	N/A	N/A	89	85	84.8	N/A	80.2	N/A	82.45

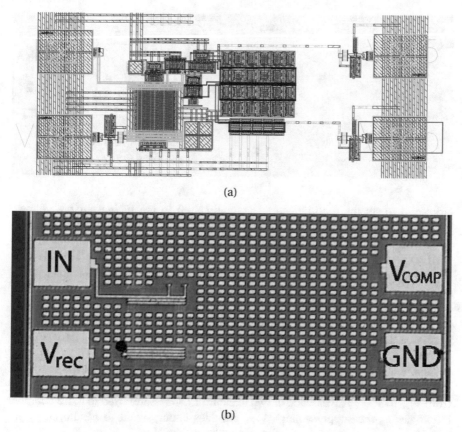

(a)

(b)

Fig. 5.38 (a) Layout and (b) photomicrograph of fabricated chip in 0.18 μm HV CMOS process. The PAD size is 115 μm × 140 μm and the space between PADs is 100 μm

A novel synchronous rectifier was proposed and compared to the state of the art as shown in Fig. 5.38. The rectifier was designed in 0.18 μm HV CMOS process, and measured along with simulated results were shown in good agreement.

However, the LIPON battery [34] requires a constant voltage of 4.1 V to be recharged, to this end a linear regulator is implemented within the same technology. Hence, the design of a two-stage low-drop-out (LDO) regulator is discussed next.

5.5 Voltage Regulator

The Infinite Power Solutions (IPS) thin film battery [34] can be recharged with constant voltage (CV) without exceeding the above limit of 4.15 V. Moreover, the typical value to charge an IPS battery is 4.1 V (V_{BATT}), so 1% variation is allowed. To this end, a regulator with low-power supply rejection ratio (PSRR) may be

Fig. 5.39 Low-drop-out
(LDO) regulator block
diagram

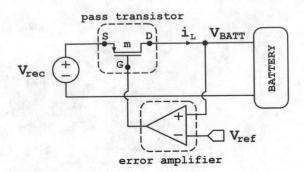

employed to set the rectifier output voltage (V_{rec}) to the typical charge voltage.
For battery operated medical implants, a low-drop-out (LDO) regulator is usually
employed to charge a battery since offers good PSRR at high frequency (above
1 MHz), does not require a controlling-driver and occupies small chip area [35–38].
Figure 5.39 is a simplified block diagram of the low-drop-out regulator made up of
a pass transistor m and an error amplifier which helps to keep V_{BATT} equal to V_{ref}.

The LDO output is always lower than its input voltage so V_{rec} has to be higher
than $V_{BATT} + V_{SD}$, where V_{SD} is the source-to-drain voltage across the pass
transistor. Therefore, the power efficiency can be poor in case $V_{rec} \gg V_{BATT}$
as this cause a large voltage drop across the pass transistor.

The regulator is designed in 0.18 μm HV CMOS technology as the output of
the rectifier can be easily higher than 4.1 V (see Fig. 5.34). To improve the PSRR
from V_{rec} to V_{BATT}, a cascade of two standard LDOs is preferred to more complex
topologies as the supply voltage V_{rec} provides enough voltage headroom to the
transistors.

5.5.1 Design of a Two-Stage Low-Drop-Out Regulator

A cascade of two low-drop-out regulators is used within the implanted device to
reduce rectifier ripple and to set the charge voltage for the micro-energy cell to
4.1 V. The LDOs are implemented in XFAB XH018 CMOS technology. Figure 5.40
depicts the linear DC-to-DC converter that sets the charging voltage of the battery.
A current mirror OTA (OTA1) and a basic OTA (OTA2) are used to enhance PSRR.
The rectifier output voltage (V_{rec}) is regulated to $V_{OUT1} = 4.2$ V and $V_{BATT} =$
4.1 V. To evaluate PSRR with respect to V_{rec}, the reference voltage (V_{ref}) is
provided externally and set to 1.2 V.

The micro-energy cell is represented by its Thevenin equivalent model with
an equivalent capacitance of 277 mF (C_{BATT}), a cell equivalent series resistance
of 180 Ω (ESR), and a self-discharge resistor (R_{self}) of 10 MΩ representing the
battery leakage.

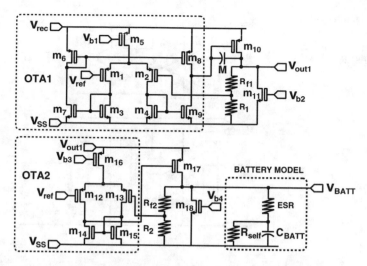

Fig. 5.40 Linear DC-to-DC converter to improve PSRR

5.5.2 PSRR Analysis

A methodology to analyze PSRR is proposed in [39] by mean of control-theory. Figure 5.41 represents an equivalent network of the two-series LDO to analyze PSRR at low frequency. Hence, V_{OUT1} and V_{BATT} can be expressed as follows:

$$V_{OUT1} = \frac{A_1 A_2}{1 + \beta_1 A_1 A_2} V_{ref} + \frac{A_2 (1 - A_{p1}) + A_{p2}}{1 + \beta_1 A_1 A_2} V_{rec} \qquad (5.21a)$$

$$V_{BATT} = \frac{A_3 A_4}{1 + \beta_2 A_3 A_4} V_{ref} + \frac{A_4 (1 - A_{p3}) + A_{p4}}{1 + \beta_2 A_3 A_4} V_{OUT1} \qquad (5.21b)$$

where the attenuation factors are defined as $\beta_1 = \frac{R_1}{R_1 + R_{f1}}$ and $\beta_2 = \frac{R_2}{R_2 + R_{f2}}$. The supply gains are written as $A_{p1} = V_1 / V_{rec}$, $A_{p2} = V_{OUT1} / V_{rec}$, $A_{p3} = V_3 / V_{OUT1}$, and $A_{p4} = V_{BATT} / V_{OUT1}$. So, assuming $\beta_1 A_1 A_2 \gg 1$ and $\beta_2 A_3 A_4 \gg 1$, Eqs. (5.21) can be expressed as:

$$V_{OUT1} \approx \frac{V_{ref}}{\beta_1} + \frac{V_{rec}}{\beta_1} \frac{1}{PSRR_{LDO_1}} \qquad (5.22a)$$

$$V_{BATT} \approx \frac{V_{ref}}{\beta_2} + \frac{V_{OUT1}}{\beta_2} \frac{1}{PSRR_{LDO_2}} \qquad (5.22b)$$

Fig. 5.41 Equivalent network of the linear DC-to-DC converter to improve PSRR

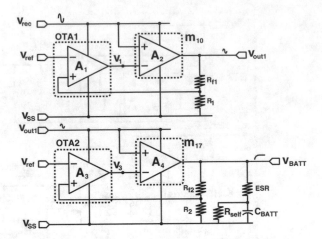

where $\frac{1}{PSRR_{LDO_1}} = \frac{1}{A_1} + \frac{1}{A_1 PSRR_2} + \frac{1}{PSRR_1}$ and $\frac{1}{PSRR_{LDO_2}} = \frac{1}{A_3} + \frac{1}{A_3 PSRR_4} +$ $\frac{1}{PSRR_3}$. The parameters $PSRR_i$ are defined by $PSRR_1 = -\frac{A_1}{A_{p1}}$, $PSRR_2 = \frac{A_2}{A_{p2}}$, $PSRR_3 = -\frac{A_3}{A_{p3}}$, and $PSRR_4 = \frac{A_4}{A_{p4}}$. By replacing Eq. (5.22a) into Eq. (5.22b), V_{BATT} can be expressed as follows:

$$V_{BATT} = \frac{1}{\beta_2} \left[\left(1 + \frac{1}{\beta_1 PSRR_{LDO_2}} \right) V_{ref} + \left(\frac{1}{PSRR_{LDO_1} PSRR_{LDO_2}} \right) \frac{V_{rec}}{\beta_1} \right]$$

(5.23)

At low frequency any variation in V_{rec} is further suppressed at V_{BATT} by using two LDOs in series as shown by the term $(PSRR_{LDO_1} \times PSRR_{LDO_2})^{-1}$.

5.5.3 Experimental Results

PSRR of the proposed LDO is measured using the methodology proposed in [40]. At low-frequency PSRR measured at V_{BATT} is 10 dB above PSRR measured at V_{OUT1} (Fig. 5.42a), while at high frequency both PSRRs dropped down to few dBs (Fig. 5.42b). To compensate for the very low-frequency pole introduced by the micro-cell battery and to improve PSRR at high frequency, a parallel bypass capacitor with low ESR can be added (Fig. 5.42b).

Table 5.5 is a summary of the measured performance of the proposed LDO. Measured results are very close to simulated results (typical case), the variation is due to a shift in the pmos threshold voltage from -1.4 V (typical) to -1.2 V as reported by the process control monitor (PCM) characterization report of the foundry.

Fig. 5.42 Measured PSRR: (**a**) at low frequency from 50 Hz to 1 kHz and (**b**) at high frequency from 10 kHz to 1 MHz

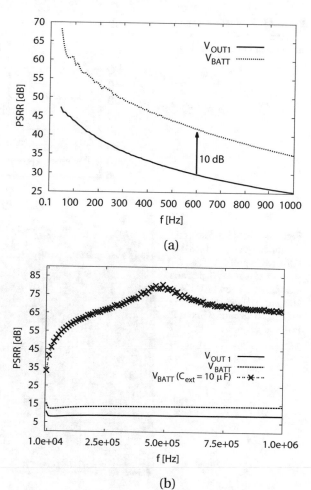

(a)

(b)

Table 5.5 Summary of LDO specifications and performance

Parameter	
Input voltage (V)	0–9 (V_{rec})
Load current (mA)	10 (i_{OUT})
Output voltage (V)	4.1 ± 0.05 (V_{BATT})
Battery capacity (μAh)	300 [34]
Drop voltage (mV)	200 ($= V_{rec} - V_{BATT}$)
Ground current (μA)	210
Load regulation @ V_{OUT1} (mV/mA)	0.66 (simulated 0.6)
Line regulation $\frac{\Delta V_{OUT1}}{\Delta V_{rec}}$ (mV/V)	2.2 (simulated 0.8)
Load regulation @ V_{BATT} (mV/mA)	4.7 (simulated 1.6)
Line regulation $\frac{\Delta V_{BATT}}{\Delta V_{rec}}$ (mV/V)	0.08 (simulated 0.03)

Fig. 5.43 Regulator power conversion efficiency as a function of the output current

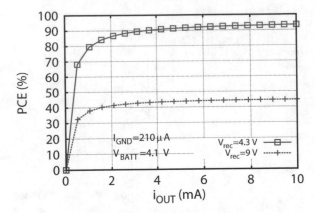

Fig. 5.44 Photomicrograph of fabricated chip in 0.18 μm HV CMOS process. The PAD size is 115 μm × 140 μm and the space between PADs is 100 μm. The total active area measure 0.48 mm² without PADs

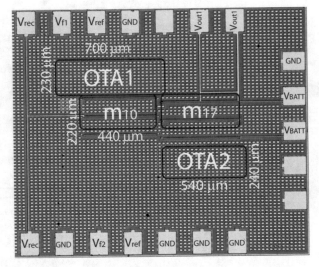

The power conversion efficiency of a linear regulator can be expressed as

$$\eta_{reg} = \frac{V_{BATT} i_{OUT}}{V_{rec} (I_{GND} + i_{OUT})} \tag{5.24}$$

where I_{GND} represents the ground current of the regulator. Figure 5.43 shows the power conversion efficiency of the proposed regulator for two different input voltages V_{rec} equal to 4.3 and 9 V. For the latter case, the efficiency results to be quite low since higher than the output voltage.

Figure 5.44 shows the photomicrograph of the LDO in 0.18 μm HV CMOS process. The PAD size is 115 μm × 140 μm and the space between PADs is 100 μm. The total active area measure 0.48 mm² without PADs. The width and length of the pass transistors m_{10} and m_{17} are 10.8 mm/2.9 μm, feedback resistors R_1 and R_2 are

$100\,k\Omega$, the differential pairs m_1–m_2 and m_{12}–m_{13} are biased in weak-inversion to reduce the offset due to device mismatch.

In the following subsections, details about the wireless communication are discussed.

5.6 OOK/ASK Demodulator

The need of continuous monitoring medical applications along with advances in low-power electronics led to the development of sensor nodes (medical implants) for ubiquitous healthcare system [or wireless body area networks (WBANs)] [41]. The primary functions of these sensor nodes are to unobtrusively sample vital signs such as heart rate, blood pressure, temperature, oxygen saturation, etc. As a result, people who are more fragile to health diseases (e.g. elderly patients) can be treated in time in case of an emergency as well as reduce the number of visits to doctors.

The WBAN is configured and controlled by a personal server (or control unit) that gives an identification (ID) number to each sensor node through wireless communication (or forward link). After initialization, the relevant data can be transferred from a sensor to the personal server (or return link) thus an addressing procedure is used where only the sensor that contains the corresponding ID wakes up [42].

An inductive-telemetry system is normally implemented for implantable cardiac pacemakers, and low-power amplitude-shift keying (ASK) modulation schemes have been proposed [43–45]. Magnetic links used in transdermal applications are energy inefficient for deep implanted medical device (IMD), wherein ultrasound waves have been shown as valid alternative [46].

Since deep implanted sensor nodes are impractical to replace the battery, a rechargeable battery is required. To limit the number of recharging cycles and to preserve the autonomy of the rechargeable μbattery, a low-power receiver architecture needs to be envisaged. Based on the communication protocol and energy harvesting strategy, the receiver can be always on or activated only when necessary via a wake-up receiver (WuRX) [47, 48].

Figure 5.45 shows the operational principle of an ultrasound health monitoring system during the addressing mode. A pressure field is generated through a transducer connected to a power amplifier which allows data transmission; four sensor nodes are listening the message that has to be demodulated.

Figure 5.46 shows the block diagram of the receiver architecture in the sensor node. To obtain a low-power receiver, an OOK/ASK modulation scheme is employed since removes the use of a local oscillator and synthesizer. Moreover, the use of ultrasound in the medical therapeutic range (from 1 to 3 MHz) for data communications in wireless sensor nodes reduce notably the power consumption as the circuit speed decreases. The demodulator is connected directly to the piezoelectric

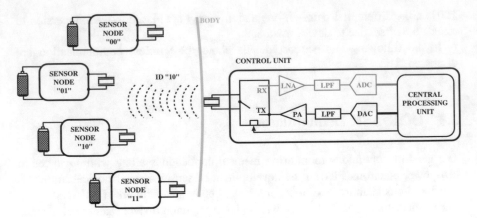

Fig. 5.45 Health monitoring system operational principle

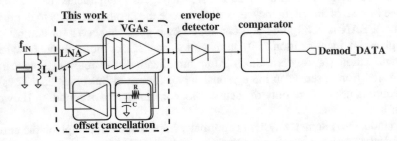

Fig. 5.46 Block diagram for OOK modulated ultrasound signals

transducer in parallel with an inductor (L_P) used as resonance matching network to increase the receiver sensitivity. The rest of the receiver is made up of a low-noise-amplifier (LNA), a cascade of variable-gain-amplifiers (VGAs), an envelope detector, and a hysteresis comparator. A high-pass RC filter is used as DC-offset cancelation technique to compensate the offset voltage due to device mismatch [49].

The design of low-power gain stages in standard $0.18\,\mu m$ CMOS technology is presented next.

5.6.1 Quasi-Identical Cascaded Gain Stages

The front-end block of the proposed receiver architecture is based on first-order n-stages quasi-identical differential amplifiers. However, the gain stages are assumed to be identical for simplicity of the analysis.

The gain-bandwidth (GBW) product of the unit gain cell for m identical cascaded stages is given by [50]

$$GBW_{cell} = \frac{BW_{tot}}{\sqrt{\sqrt[m]{2} - 1}} \sqrt[m]{A_{tot}} \qquad (5.25)$$

where $BW_{tot}/\sqrt{\sqrt[m]{2}-1}$ and $\sqrt[m]{A_{tot}}$ represent the bandwidth and gain of the unity gain cell. An ultrasound transducer can yield an output voltage equal to (see Sect. 5.1.2)

$$V_P = \frac{I_P}{\omega C_P} \qquad (5.26)$$

With a minimum received signal current $I_P = 500\,\text{nA}$ at 1 MHz, and a plate capacitance $C_P = 260\,\text{pF}$, the output peak voltage V_P is equal to 306 μV. Hence, a total gain A_{tot} of 58 dB is required to obtain a peak-to-peak voltage of 500 mV at the output of m gain stages.

Figure 5.47a plots the cell GBW as a function of the number of gain stages for different total BW. Figure 5.47b shows the gain cell versus the number of gain stages to obtain a total gain of 58 dB. For m lower than three the GBW is above 30 MHz and the gain per stage is above 15 dB. While for m above four the GBW stays above 10 MHz and the gain per stage decreases to a minimum of 6 dB at $m = 10$. However, as the number of gain stages increases the die area increases as well, the number of amplifying stages can be set between four and six.

Depicted in Fig. 5.48 is the fully differential amplifier with resistive load R_L, used as basic building block in the cascaded gain stages structure. To achieve low-power consumption the input stage (m_x) is sized in weak-inversion, thus the single-ended gain can be expressed as

$$A_O = \frac{1}{2} g_{m,x} R_L = \frac{I_B/2}{2nU_T} \times \frac{(V_{DD} - V_{OUT,p})}{I_B/2} = 8I_B R_L \qquad (5.27)$$

where $g_{m,x}$ represents the transconductance of the input stage, I_B the biasing current, $n = 1.2$ the slope factor, and U_T the thermal voltage (26 mV at 300 K). The coefficient 8 is expressed in V^{-1} and is derived as $(4nU_T)^{-1}$.

The unit cell GBW can be easily determined as

$$GBW_{cell} = \frac{8I_B}{C_L} \qquad (5.28)$$

where C_L represents the input capacitance of the next stage, thus the bandwidth BW_{cell} is equal to $1/R_L C_L$. Combining (5.28) and (5.25), the unit cell biasing current can be written as

$$I_{B,cell} = \frac{C_L}{8} \frac{BW_{tot} \sqrt[m]{A_{tot}}}{\sqrt{\sqrt[m]{2} - 1}} \qquad (5.29)$$

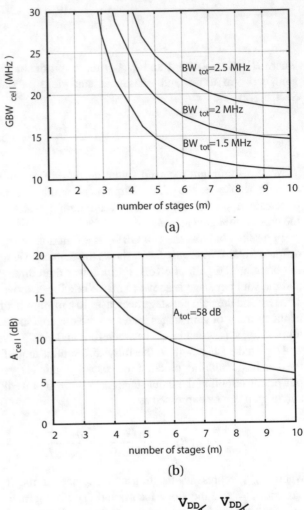

Fig. 5.47 (**a**) Gain-bandwidth product versus number of gain stages. (**b**) Unit gain cell versus number of gain stages

(a)

(b)

Fig. 5.48 Differential pair circuit with passive loads

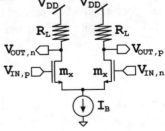

Figure 5.49 shows the unit cell biasing current as a function of number of gain stages for three load capacitor values at $BW_{tot} = 1.5$ MHz and $A_{tot} = 58$ dB. As the number of stages increases the current per unit cell saturates, thus for low-frequency operation increasing the number of cells degrades the noise performance.

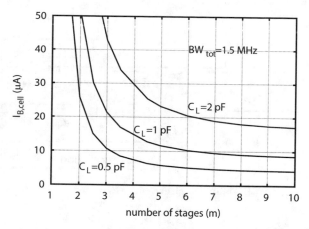

Fig. 5.49 Biasing current versus number of gain stages in weak-inversion for a $BW_{tot} = 1.5\,\text{MHz}$

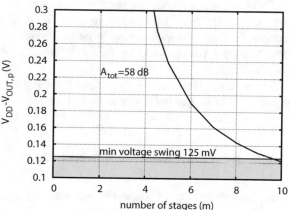

Fig. 5.50 DC voltage drop at the output of a unit cell gain

To complete the analysis the load resistor can be expressed as

$$R_L = \frac{\sqrt{\sqrt[m]{2} - 1}}{BW_{tot}C_L} \tag{5.30}$$

For $BW_{tot} = 1.5\,\text{MHz}$ and $C_L = 0.5\,\text{pF}$, R_L results tremendously large such that the thermal noise increases as directly proportional to the resistor value.

To limit the power consumption of m gain stages, the power supply can be reduced to some extent before the output voltage saturates due to high-voltage gain. The DC output voltage for m gain stages is derived from (5.27)

$$V_{DD} - V_{OUT,p} = \frac{\sqrt[m]{A_{tot}}}{16} \tag{5.31}$$

Figure 5.50 plots the DC voltage drop at the output of a unit cell gain. For m lower than four, additional DC biasing is required for each input pair.

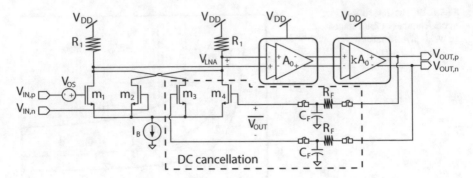

Fig. 5.51 DC-offset cancelation circuit technique

To assure at least a peak-to-peak voltage swing of $\pm 125\,\text{mV}$, the supply voltage can be lowered from $1.8\,\text{V}$ down to $1.5\,\text{V}$ in standard $0.18\,\mu\text{m}$ CMOS process. Finally, four gain stages are used to build the front-end gain amplifier. Moreover, the DC output voltage yield by each amplifier allows directly DC biasing for the following stage.

To avoid the saturation of the output voltage due to device mismatch, an offset cancelation technique is implemented and is discussed next.

5.6.2   DC-Offset Cancelation Technique

The gate voltage mismatch (σ_{V_G}) between the input pair transistors of a differential amplifier is a function of two independent statistical values: the threshold voltage mismatch (σ_{V_T}) and the current mismatch (σ_β). For MOS transistors operating in weak-inversion an input referred offset of $5\,\text{mV}$ can be obtained [51]. Hence, a DC-offset cancelation technique is required for high-gain amplifiers to avoid the saturation of the output voltage.

Figure 5.51 shows the DC-offset cancelation circuit technique used in this work. The use of a low-noise-amplifier along with a summing amplifier to remove the presence of offset voltage V_{OS} due to device mismatch is merged. The offset extraction is implemented through a low-pass filter ($\tau = R_F C_F$).

The output voltage of the amplifier can be expressed as

$$V_{OUT} = V_{OUT,p} - V_{OUT,n} = A_{tot} \times V_{LNA} \tag{5.32}$$

where V_{LNA} represents the output of the compensation circuit and can be found as

$$V_{LNA} = g_m R_1 \left[(v_{IN} + V_{OS}) - \overline{V_{OUT}} \right] \tag{5.33}$$

where g_m represents the transconductance of m_1 and m_2 transistors and R_1 the load resistor. From (5.32) and (5.33), the DC component can be expressed as

$$\overline{V_{LNA}} = g_m R_1 \left(V_{OS} - \overline{V_{OUT}}\right) \tag{5.34a}$$

$$\overline{V_{OUT}} = A_{tot} \overline{V_{LNA}} \tag{5.34b}$$

and the AC component is

$$v_{LNA} = g_m R_1 v_{IN} \tag{5.35a}$$

$$v_{OUT} = A_{tot} v_{LNA} \tag{5.35b}$$

Hence, combining (5.34a) and (5.34b) the DC output voltage can be seen as

$$\overline{V_{OUT}} = \frac{g_m R_1 A_{tot} V_{OS}}{1 + g_m R_1 A_{tot}} \approx V_{OS} \tag{5.36}$$

and combining (5.35a) and (5.35b) the AC output voltage as

$$v_{OUT} = g_m R_1 A_{tot} v_{IN} \tag{5.37}$$

The drawback of this technique is the low-frequency pole introduced by $R_F C_F$ which limit their implementation on-chip.

5.6.3 Variable Gain Amplifier Structure

However, the amplitude of the input signal V_{IN} can rise up to 6 mV, to avoid the output from being distorted the source degeneration technique is implemented to vary the gain of a unit cell.

Figure 5.52 shows four stages gain cells used in the receiver chain. The LNA is followed by two fixed-gain stages (A_O) of 12 dB and two variable gain stages with a gain tunable (kA_O) from −4 to 10 dB. The maximum total front-end gain at 1 MHz is 60 dB.

Fig. 5.52 Four stages gain cells

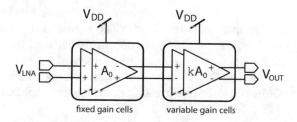

fixed gain cells   variable gain cells

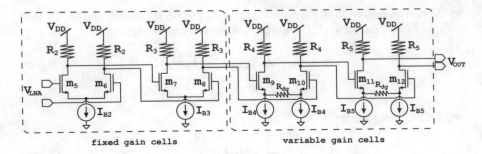

Fig. 5.53 Four gain cells architectures

Fig. 5.54 Schematic view of
the programmable
degeneration resistor

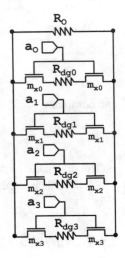

Figure 5.53 shows the implementation of the four stages using CMOS amplifier
cells with resistive load. The gain of the amplifier is tuned via the degeneration
resistance R_{dg} which is digitally controlled.

The gain of a unit variable gain cell can be written as [52]

$$kA_{4,5} = \frac{2R_{4,5}}{R_{dg} + \frac{2}{g_m}} \approx \frac{2R_{4,5}}{R_{dg}} \tag{5.38}$$

A programmable degeneration resistor is implemented using poly-resistors and
CMOS transistors biased in triode region to achieve moderate linearity and low
area consumption [53]. By default the input pins a_0 to a_3 are low, thus the gain
of the amplifier is minimum and imposed by R_0. To increase the amplifier gain,
the resistor R_0 has to be decreased, thus four resistors are added in parallel. The
equivalent resistance in each parallel branch is splitted as $R_{eq,i} = R_{dg,i} + 2R_{ON,m_{xi}}$
(Fig. 5.54).

Table 5.6 Digital input setting to control the amplifier gain

a_0	a_1	a_2	a_3	kA_0 (dB)
0	0	0	0	−4
1	0	0	0	0
1	1	0	0	4
1	1	1	0	8
1	1	1	1	10

Fig. 5.55 Schematic view of the output buffer

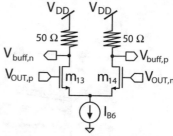

Table 5.6 shows the logic table to achieve the required voltage gain. The load resistors $R_{4,5}$ are equal to $25\,k\Omega$, and the degeneration resistor values are chosen according to (5.38).

To measure the performances of the amplifier chains an output buffer is included in the design since the test equipments are characterized with $50\,\Omega$ load. However, the output buffer can be removed from the receiver chain since unnecessary.

5.6.4   Output Buffer

The output buffer is used to drive the testing instruments with an input impedance of $50\,\Omega$. Figure 5.55 shows the differential amplifier with passive load used as output buffer since the overall bandwidth falls below $10\,MHz$. This time, the transistor pair m_{13} and m_{14} are biased in strong inversion as the biasing current is set to $10\,mA$. Moreover, the output buffer adds $2\,dB$ gain in the chain of gain stages and is taken into account in the measurements results.

Table 5.7 summarizes the design variable as biasing currents and transistor dimensions for the chain of amplifier stages. To reduce flicker noise very large input device sizes (m_1–m_4) are used in combination with resistive loads. To reduce the power consumption all transistors are biased in weak-inversion due to the low speed requirement imposed by the ultrasound. To reduce the noise contribution of each input stage the minimum channel length is increased [54].

The measurement along with simulation results are given in the next section and the measurement set-up is described which includes the output buffer.

Table 5.7 Design variables

(a) Biasing currents		(b) Transistors size			
Current	mA	Transistor	W (μm)	L (μm)	W/L
I_{B1}	0.04	m_1–m_2	60	0.5	120
I_{B2}–I_{B3}	0.016	m_3–m_4	60	0.5	120
I_{B4}–I_{B5}	0.016	m_5–m_6	48	1	48
I_{B6}	10	m_7–m_8	48	1	48
		m_9–m_{10}	48	1	48
		m_{11}–m_{12}	48	1	48
		m_{13}–m_{14}	24	0.18	133

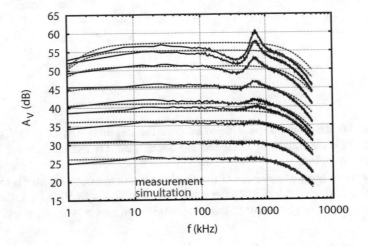

Fig. 5.56 Single-ended amplifier gain: measurement vs simulation results

5.6.5   Measurement Results

The Agilent network/spectrum analyzer 4395A is used to characterize the frequency response of the fully differential VGA. The input and output voltages of the amplifier are measured through the Tektronix active probe P6243 together with the power supply Tektronix 1103. The evaluation board from Texas Instrument THS3120, used to characterize the power supply rejection ratio (PSRR) of the low-drop-out (LDO) regulator, provides the input AC plus DC biasing voltage for the amplifier.

Figure 5.56 shows the measured versus the simulated single-ended voltage gain in dB from 1 kHz to 5 MHz. From 25 dB up to 40 dB voltage gain, the simulated results fit quite well the measurements. However, at higher voltage gain a variation of ± 2 dB is measured around 500 kHz. The cause of this mismatch is attribute to the measurement set-up as the amplifier design shows clearly a dominant pole at 2.5 MHz which determines the overall bandwidth (BW). In the measurement results the output buffer gain is included as well as in the simulation results. The gain of the output buffer is extracted from the simulation and is equal to 2 dB. The default gain is equal to 26 dB and is increased to a maximum of 56.5 dB for a total power consumption of 178 μW.

To measure the noise of the amplifier the inputs are short circuited to ground and the outputs are connected through the Mini-Circuits power combiner ZFSCJ-2-1 to the Agilent network/spectrum analyzer 4395A. Figure 5.57 shows the output referred noise measurements for all gains from 1 kHz to 1.5 MHz. The maximum noise is recorded by setting the maximum gain and is higher than $80\,\mu V/\sqrt{Hz}$ at 3 kHz.

To compute the input referred noise, the differential gain of the amplifier is measured at the fixed frequency of 1 MHz. Table 5.8 gives the simulated and measured amplifier differential gains. The measured differential gain achieves a maximum of 4 dB discrepancy from the simulated gain. This difference is acceptable as the total current to achieve almost 60 dB gain for an overall bandwidth of 2.5 MHz is only 119 μA.

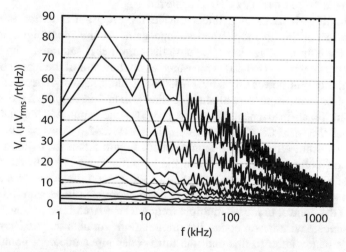

Fig. 5.57 Measured output referred noise

Table 5.8 Differential gain at 1 MHz: simulated vs. measured results

A_v (dB) simulated	A_v (dB) measured
30	28.97
34.6	34.01
40.15	38.16
44.38	42.14
46.4	44.22
50.9	48.45
56.45	52.68
60.69	56.23
62.7	58.8

Table 5.9 Output and input referred noise at 1 MHz: simulated and measured results

$V_{OUT,n}$ (μV/$\sqrt{\text{Hz}}$)		$V_{IN,n}$ (nV/$\sqrt{\text{Hz}}$)	
Sim.	Meas.	Sim.	Meas.
0.52	0.45	16.4	15.8
0.87	0.69	16.2	13.8
1.64	1.18	16.1	14.6
2.67	1.89	16.1	14.7
3.37	2.44	16.1	15.1
5.66	4.46	16.1	16.9
10.68	8.03	16.1	18.7
17.29	11.9	15.9	18.4
22.01	15.7	16.1	17.9

Hence, the input referred noise at 1 MHz can be computed as the output referred noise into the voltage gain. Table 5.9 gives the output referred noise results at different gains. The measured results are close to the simulations, and an average input referred noise of 16 nV/$\sqrt{\text{Hz}}$ is achieved.

The amplifier behaves as a linear device for a certain range of input signal amplitudes according to the defined voltage gain. As the input signal amplitude increases the output voltage increases until the amplifier is saturated. Therefore, the 1 dB compression point (P-1dB) is measured as an indicator of the input power level at which the output power drops by 1 dB. The Agilent network/spectrum analyzer 4395A is used to measure the 1 dB compression point of the amplifier, the evaluation board from Texas Instrument THS3120 provides the input AC plus DC biasing voltage and the Mini-Circuits power combiner ZFSCJ-2-1 converts the output of the amplifier from differential to single-ended. Figure 5.58a shows the input 1 dB compression point is 42.08 dBm when the amplifier gain is set to the minimum voltage gain (28.97 dB). Figure 5.58b shows the input 1 dB compression point is 69.9 dBm when the amplifier gain is set to the maximum voltage gain (58.8 dB).

In case the control unit uses multiple frequencies (f_1 and f_2) to transfer energy or communicate with sensor nodes, intermodulation products are yielded within the amplifier's bandwidth. To this end, the third-order input intercept point (IIP3) is used to characterize this effect. Two signals at frequency $f_1 = 1$ MHz and $f_2 = 1.03$ MHz are fed into the Mini-Circuits power combiner ZFSCJ-2-1 whose output is connected to the amplifier input. A second Mini-Circuits power combiner ZFSCJ-2-1 converts the output of the amplifier from differential to single-ended which is connected to the Agilent network/spectrum analyzer 4395A that measures the signal power at f_1, f_2, $2f_1 - f_2$, and $2f_2 - f_1$. A graph is plotted between the output power versus the input power. Figure 5.59 shows the linearly amplified signal at f_1 and the intermodulation product $2f_1 - f_2$. Both curves are extrapolated and the input power at which the intermodulation product curve intersects the linearly amplified signal curve is the IIP3 of the amplifier. From Fig. 5.59a the IIP3 of the VGA at 28.97 dB gain is -30.85 dBm. While from Fig. 5.59b the IIP3 of the VGA at 58.8 dB gain is -52.17 dBm.

Fig. 5.58 P_{OUT} versus P_{IN}
curve for P-1dB measurement
with: (**a**) minimum- and (**b**)
maximum-VGA gain

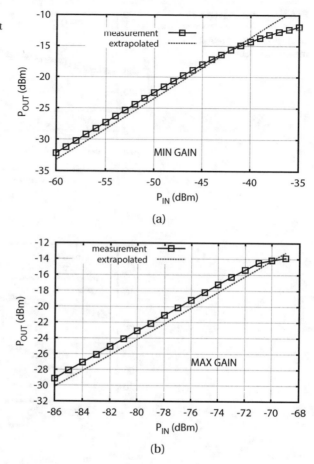

(a)

(b)

The measured results of the quasi-identical cascaded gain stages are presented. The measured single-ended voltage gain versus frequency is presented along with simulated results. The output referred noise is measured and compared with simulation results. Lastly, 1 dB compression point and the third-order input intercept point measurement at different amplifier gain settings are shown.

5.6.6   Summary

This section explained the idea behind a low-power non-coherent receiver in wireless body area network for sensor nodes wherein the power usage is limited and dictated by the life-time of the energy storage element. To this end a low-power front-end circuit is conceived which is aimed to demodulate the incoming message from the control unit during the addressing phase.

Fig. 5.59 P_{OUT} versus P_{IN}
curve for IIP3 measurement
with: (**a**) minimum- and (**b**)
maximum-VGA gain

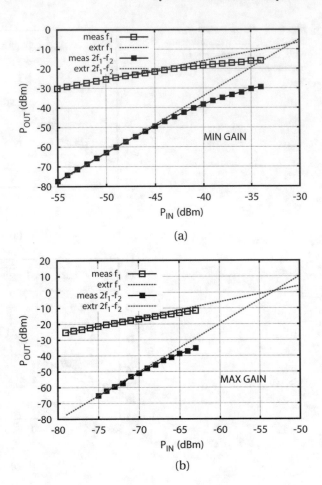

(a)

(b)

The front-end circuit of the receiver consists of a low-noise-amplifier, two fixed-gain stages, and two variable gain stages. The cascaded gain stages amplifier is designed and fabricated in standard CMOS technology.

Figure 5.60 shows the photomicrograph of the VGA in 0.18 μm HV CMOS process. The PAD size is 115 μm × 140 μm and the space between PADs is 100 μm. The total active area measure 0.081 mm^2 without PADs.

The cascaded gain stages amplifier is characterized and a differential gain of 56.8 dB at 1 MHz without the output buffer is shown. The power consumption of the amplifier is only 178 μW showing a total bandwidth of 2.5 MHz.

In the next section, the return link communication is discussed.

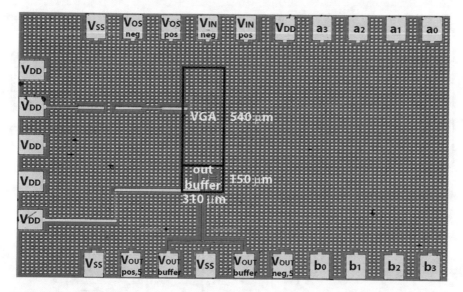

Fig. 5.60 Photomicrograph of fabricated chip in 0.18 μm HV CMOS process. The PAD size is 115 μm × 140 μm and the space between PADs is 100 μm. The total active area measure 0.081 mm² without PADs and output buffer

5.7   Load Shift Keying

A sensor node after being initialized and addressed, wakes up and starts sending the acquired data as patient's vital signs to the personal server. The backscattering technique or load modulation is used to establish a communication between the base station and the implanted device. This principle is based on the reflection of the ultrasound waves, the same modulation technique is well known in RFID systems [55, 56]. It takes advantage of the reflection coefficient variation at the interface between the ultrasound transducer and the sensor node input circuit.

5.7.1   Impedance Modulation

Figure 5.61 describes the return data communication through the backscattering principle. The control unit generates an acoustic power P_{ac} which is transmitted towards the piezoelectric transducer. The transponder modulates the incident acoustic power $P_{ac,i}$ by changing the transducer load. The incident energy can be divided into absorbed $P_{ac,t}$ and reflected $P_{ac,r}$ acoustic power. One way to have a reliable communication link is to maximize the difference between $P_{ac,t}$ and $P_{ac,r}$, thus an ASK modulation scheme is obtained. However, by varying the reflection coefficient Γ at the input of the implanted circuit phase variation also occurs between $P_{ac,i}$ and $P_{ac,r}$, thus a phase-shift-keying (PSK) modulation scheme is obtained.

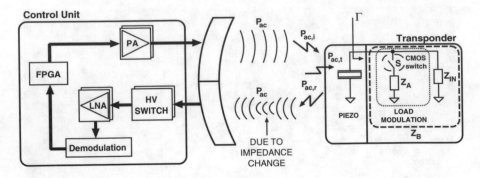

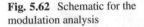

Fig. 5.61 Backscattering technique

Fig. 5.62 Schematic for the
modulation analysis

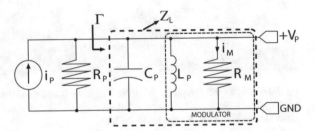

As the system is battery-powered, the AC-to-DC converter is disconnected
during the load modulation so $Z_{IN} \rightarrow \infty$. Hence, the input impedance of the
ultrasound sensor node is mainly influenced by the modulator represented by the
series connection between an impedance Z_A and a CMOS switch.

Figure 5.62 shows the equivalent model of the piezoelectric transducer during the
modulation. The modulator can be modeled as a simple resistor R_M in parallel with
the transducer made up of a current source which peak value is $\widehat{i}_P$, a resistor R_P,
and to cancel out the capacitor C_P at 1 MHz a parallel inductor L_P is added. Hence,
an ASK modulator is used since the reactive part are negligible. If R_M is equal to
$0\,\Omega$, the transducer is short circuited to ground thus preventing it to vibrate and to
absorb power. Consequently, the reflection coefficient module is equal to 1 and total
power reflection occurs. Conversely when R_M is equal to R_P, the available power
from the transducer is ideally completely absorbed by R_M if power-match occurs.

The current through the modulator resistance can be written as

$$\widehat{i}_M = \widehat{i}_P \alpha \frac{\sqrt{1 + Q_M^2}}{\sqrt{(1 + \alpha)^2 + (\alpha Q_M - Q_P)^2}} \tag{5.39}$$

where α represents R_P/R_M and $Q_M = R_M/\omega L_P$, $Q_P = \omega C_P R_P$ the quality
factors. Power-match occurs when $\alpha = 1$ and $\omega^2 C_P L_P = 1$. Therefore, (5.39)
becomes

$$\widehat{i}_{M,max} = \widehat{i}_P \frac{\sqrt{1 + Q_M^2}}{2} \tag{5.40}$$

and for large value of Q_M, $\widehat{i}_M = \widehat{i}_P Q_M/2$. $\widehat{i}_M$ normalized to its maximum value (5.40) is equal to

$$\widehat{i}_{M,norm} = \frac{\widehat{i}_M}{\widehat{i}_{M,max}} = \frac{2\alpha}{\sqrt{(1 + \alpha)^2 + (\alpha Q_M - Q_P)^2}} \tag{5.41}$$

for $\omega^2 C_P L_P = 1$, (5.41) can be expressed as

$$\widehat{i}_{M,norm} = \frac{2\alpha}{\sqrt{(1 + \alpha)^2 + Q_M^2 (\alpha - 1)^2}} \tag{5.42}$$

Figure 5.63 plots the normalized input current $i_{M,norm}$ versus α at different quality factor values. As the quality factor increases, $i_{M,norm}$ gets more sensitive to α.

Hence, the quality factor Q_M plays an important role in terms of energy delivered to the modulator resistance R_M. Indeed, the current through the modulator is lower respect to its maximum for $\alpha \neq 1$, thus the generator current i_P is consumed mostly by R_P which radiate this power back to the control unit. To analyze the amount of power that is backscattered to the control unit, the reflection coefficient Γ is expressed as

$$\Gamma = \frac{Z_L - R_P}{Z_L + R_P} \tag{5.43}$$

Fig. 5.63 Normalized input current $i_{M,norm}$ versus α

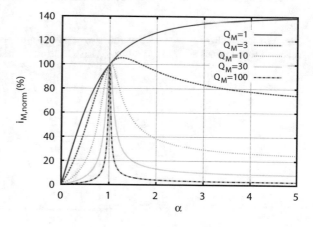

where $Z_L = \frac{j\omega L_P R_M}{R_M + j\omega L_P - \omega^2 L_P C_P R_M}$, with $Q_M = Q_P$ and $\alpha = R_P/R_M$, thus (5.43) can be written as follows

$$\Gamma = \frac{Q_M (1 - \alpha) + j (1 - \alpha)}{Q_M (\alpha - 1) + j (1 + \alpha)} \qquad (5.44)$$

by adding and subtracting $2j$ at the numerator, (5.44) can be simplified as

$$\Gamma = 1 - \frac{2j}{Q_M (\alpha - 1) + j (1 + \alpha)} \qquad (5.45)$$

The amount of power that is reflected during the reflection ($P_{r,r}$) or absorption ($P_{r,a}$) state by the modulator is a function of the available power P_{AV} from the transducer. If the modulator shorts the piezoelectric transducer to ground or leaves it unloaded, the reflection coefficient Γ is equal to ∓ 1 and the reflected power in the reflection state $P_{r,r}$ is equal to P_{AV}. If the modulator terminate the piezoelectric transducer with an impedance Z_L, the reflected power during the absorption state $P_{r,a}$ is defined as

$$P_{r,a} = P_{AV} |\Gamma|^2 = P_{AV} \frac{\left(1 + Q_M^2\right) (\alpha - 1)^2}{(1 + \alpha)^2 + Q_M^2 (1 - \alpha)^2} \qquad (5.46)$$

Figure 5.64 plots the normalized reflected power versus α for different quality factors. For $\alpha \neq 1$ and high quality factor values, a small mismatch between R_P and R_M leads to a drastic increase of $P_{r,a}/P_{AV}$.

Moreover, R_M is close to a short-circuit condition if $\alpha > 1$ thus decreasing the difference between the absorption and the reflection state at the base station. Obviously, $R_M = R_P$ gives the best operating condition. Hence, the modulation depth is thus very dependent on the factor α for moderately high quality factors.

The implementation of the modulator along with simulated results are given next.

Fig. 5.64 Normalized reflected power versus α

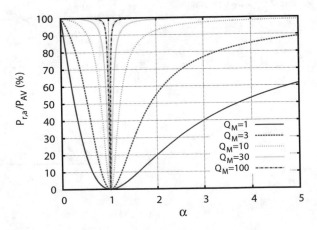

5.7.2   Modulator Architecture

A simple architecture is adopted to implement the modulator circuit, which is done in discrete components. A metal-oxide semiconductor field-effect transistor (MOSFET) is used to open or short the transducer terminals to allow ultrasound sensor data transfer for implanted medical device [57] and through-wall systems [58, 59]. Moreover, the same technique is employed to transfer information via inductive link in cortical neural interfaces [60–62].

Figure 5.65 plots the data rate versus the distance between a transmitter and receiver pair for ultrasound and inductive links.

The ultrasound link shows low data rate performance essentially due to the low operational frequency of 1 MHz, while the inductive link works at frequency above 10 MHz. As function of distance the data rate decreases as well as the transmitted energy thus lower frequencies are more favorable to transmit data from deep implanted devices to a control unit.

In this work, as the distance between the control unit and the implanted medical device is equal to 10.5 cm, which is the focal point of the spherical transducer array (Fig. 3.12b), and the working frequency is set to 1 MHz (Sect. 3.2.1), a data rate from 5 to 20 kbps is targeted for the uplink.

Figure 5.66 shows the block diagram of the impedance modulator, which is implemented through a parallel combination of a N-channel (m_1) and a P-channel (m_2) MOSFETs. To avoid that body diodes conduct, two diodes D_1 and D_2 are added in series with m_1 and m_2. The positive branch is used to clip the positive half-wave and the negative branch is used to clip the negative half-wave of the input current. Hence, m_1 and m_2 are activated/deactivated simultaneously via enable commands En_Comm_NMOS and En_Comm_PMOS generated by a μcontroller. In addition, the μcontroller switches the reference voltage from ground to half of the supply voltage during the uplink via the pin ref_ctrl.

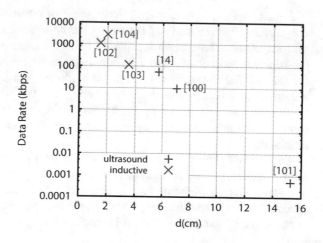

Fig. 5.65 Data rate versus the distance between a transmitter and receiver pair for ultrasound and inductive links

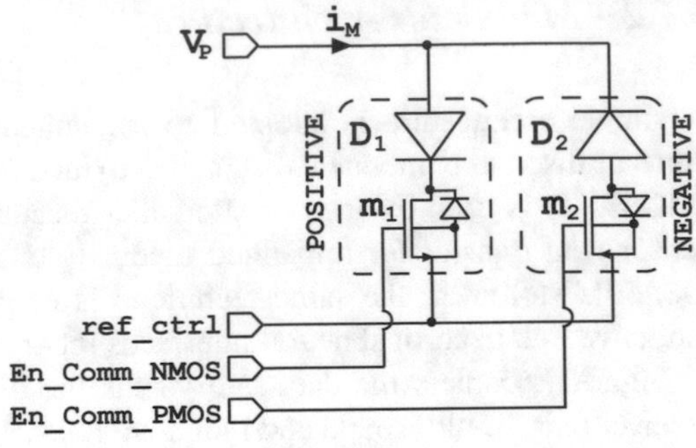

Fig. 5.66 Block diagram of
the impedance modulator

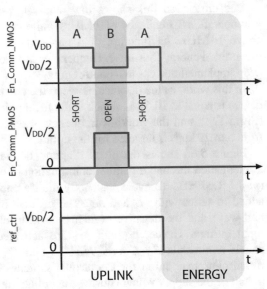

Fig. 5.67 Timing signals
during load modulation and
energy harvesting phases

Figure 5.67 plots the voltage waveforms En_Comm_NMOS, En_Comm_PMOS, and ref_ctrl during the uplink and energy harvesting phases. The impedance modulation is divided in two states: A and B. The ultrasound transducer is short circuited to $V_{DD}/2$ during the state A and is left unloaded during the state B. The total current through the modulator during the short-circuit state is non-zero since the transistors are conducting and the piezo voltage V_P is clipped to a maximum voltage set by diodes threshold voltage. While the total current through the modulator during the unloaded state is zero and the piezo voltage V_P is maximum.

Figure 5.68 shows the current i_x and voltage V_P waveforms at the modulator input terminals during the short-circuit state A and open-circuit state B. The voltage V_P during the short-circuit state is non-zero due to the diodes $D_1 - D_2$, so the reflected power is less than P_{AV}; whereas for the open-circuit state as the current i_x is equal to zero the reflected power is equal to P_{AV}.

Fig. 5.68 Current and voltage waveforms at the modulator terminals during state *A* and *B*

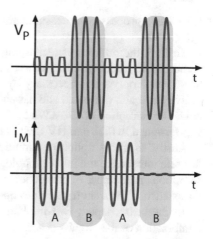

The best condition for this type of data transfer is to switch the modulator impedance from the matched condition ($R_M = R_P$) to the unmatched condition (open or short). However, as the implanted ultrasound transducer is made up of an array of elements to limit the number of transistors and control signals from the μcontroller, the open and short conditions modulation is adopted.

5.8 Summary

This chapter presented the implanted medical device architecture; the energy harvesting and the wireless communication circuits are described. First, an equivalent model of the ultrasound piezoelectric transducer was introduced and validated through simulated and measured results. Hence, the piezoelectric transducer was modeled as a simple current source in parallel with a parallel combination of a capacitor and a resistor.

Next, the main blocks for the recharge of the implanted μbattery were shown: the rectifier, the linear regulator, and high-/low-voltage battery detection circuits to prevent damages. A design methodology for passive rectifier was proposed which uses a periodic steady state simulation to model the nonlinear input impedance of the rectifier. Then, different rectifier topologies used for ultra high frequency were designed and fabricated in 0.18 μm CMOS process. To increase the voltage conversion efficiency of the rectifier, a transistor should be used as switch rather than as a diode. Therefore, a novel synchronous rectifier was proposed and compared to the state-of-the-art works. The rectifier was designed in 0.18 μm HV CMOS process, and measured along with simulated results were shown in good agreement.

A tight constant voltage was required to recharge the μbattery, thus a two-stage low-drop-out regulator was designed to increase the power supply rejection ratio from the rectifier output to the battery input. The regulator was designed in 0.18 μm

HV CMOS process, and measured along with simulated results were shown in good agreement.

To limit the number of recharging cycles and to preserve the autonomy of the rechargeable μbattery, a low-power receiver architecture was envisaged which can be activated only when necessary by the implant. The front-end amplifier made up of five stage was presented and has shown a differential gain of 56.8 dB at 1 MHz, a power consumption of 178 μW at 1.5 V and a bandwidth of 2.5 MHz. The amplifier was designed in 0.18 μm HV CMOS process.

Lastly, to allow uplink communication from the implant to a control unit, a discrete impedance modulator topology was proposed. To minimize the power consumption and number of components, the load impedance of ultrasound transducer was switched from short-circuit to open-circuit state.

In the next chapter, two implants architecture will be presented which use a combination of the circuits described in this chapter.

References

1. J.-P. Curty, N. Joehl, C. Dehollain, M.J. Declercq, Remotely powered sddressable UHF RFID integrated system. IEEE J. Solid-State Circuits **40**(11), 2193–2202 (2005)
2. R. Krimholtz, D.A. Leedom, G.L. Matthaei, New equivalent circuits for elementary piezoelectric transducers. Electron. Lett. **6**(13), 398–399 (1970)
3. H.A.C. Tilmans, Equivalent circuit representation of electromechanical transducers: Ii. Distributed-parameter systems. J. Micromech. Microeng. **7**(4), 285 (1997)
4. K.K. Shung, M. Zippuro, Ultrasonic transducers and arrays. IEEE Eng. Med. Biol. Mag. **15**(6), 20–30 (1996)
5. S. Roundy, P.K. Wright, J.M. Rabaey, *Energy Scavenging for Wireless Sensor Networks: With Special Focus on Vibrations* (Kluwer Academic Publishers, Norwell, 2004)
6. K.M. Silay, D. Dondi, L. Larcher, M. Declercq, L. Benini, Y. Leblebici, C. Dehollain, Load optimization of an inductive power link for remote powering of biomedical implants, in *IEEE International Symposium on Circuits and Systems, 2009. ISCAS 2009*, May 2009, pp. 533–536
7. R. Barnett, S. Lazar, J. Liu, Design of multistage rectifiers with low-cost impedance matching for passive RFID tags, in *2006 IEEE Radio Frequency Integrated Circuits (RFIC) Symposium*, June 2006, 4 pp.
8. N. Tran, B. Lee, J.-W. Lee, Development of long-range UHF-band RFID tag chip using Schottky diodes in standard CMOS technology, in *2007 IEEE Radio Frequency Integrated Circuits (RFIC) Symposium*, June 2007, pp. 281–284
9. J.F. Dickson, On-chip high-voltage generation in MNOS integrated circuits using an improved voltage multiplier technique. IEEE J. Solid-State Circuits **11**(3), 374–378 (1976)
10. U. Karthaus, M. Fischer, Fully integrated passive UHF RFID transponder IC with 16.7- μW minimum RF input power. IEEE J. Solid-State Circuits **38**(10), 1602–1608 (2003)
11. T. Le, K. Mayaram, T. Fiez, Efficient far-field radio frequency energy harvesting for passively powered sensor networks. IEEE J. Solid-State Circuits **43**(5), 1287–1302 (2008)
12. B.H. Calhoun, D.C. Daly, N. Verma, D.F. Finchelstein, D.D. Wentzloff, A. Wang, S.-H. Cho, A.P. Chandrakasan, Design considerations for ultra-low energy wireless microsensor nodes. IEEE Trans. Comput. **54**(6), 727–740 (2005)

13. S. Roundy, P.K. Wright, J.M. Rabaey, *Energy Scavenging for Wireless Sensor Networks with Special Focus on Vibrations* (Springer, Berlin, 2003). ISBN: 978-1-4020-7663-3

14. M. Renaud, T. Sterken, A. Schmitz, P. Fiorini, C. Van Hoof, R. Puers, Piezoelectric harvesters and MEMS technology: fabrication, modeling and measurements, in *International Solid-State Sensors, Actuators and Microsystems Conference, 2007. TRANSDUCERS 2007*, June 2007, pp. 891–894

15. G.K. Ottman, H.F. Hofmann, A.C. Bhatt, G.A. Lesieutre, Adaptive piezoelectric energy harvesting circuit for wireless remote power supply. IEEE Trans. Power Electron. **17**(5), 669–676 (2002)

16. T.T. Le, J. Han, A. von Jouanne, K. Mayaram, T.S. Fiez, Piezoelectric micro-power generation interface circuits. IEEE J. Solid-State Circuits **41**(6), 1411–1420 (2006)

17. Y.K. Ramadass, A.P. Chandrakasan, An efficient piezoelectric energy harvesting interface circuit using a bias-flip rectifier and shared inductor. IEEE J. Solid-State Circuits **45**(1), 189–204 (2010)

18. Y.K. Ramadass, Massachusetts Institute of Technology. Dept. of Electrical Engineering, and Computer Science, *Energy Processing Circuits for Low-Power Applications* (Massachusetts Institute of Technology, Department of Electrical Engineering and Computer Science, Massachusetts, 2009)

19. M. Ghovanloo, K. Najafi, Fully integrated wideband high-current rectifiers for inductively powered devices. IEEE J. Solid-State Circuits **39**(11), 1976–1984 (2004)

20. E. Sackinger, A. Tennen, D. Shulman, B. Wani, M. Rambaud, F. Larsen, G.S. Mpschytz, A 5-V AC-powered CMOS filter-selectivity booster for POTS/ADSL splitter size reduction. IEEE J. Solid-State Circuits **41**(12), 2877–2884 (2006)

21. Y.-H. Lam, W.-H. Ki, C.-Y. Tsui, Integrated low-loss CMOS active rectifier for wirelessly powered devices. IEEE Trans. Circuits Syst. II: Express Briefs **53**(12), 1378–1382 (2006)

22. N.J. Guilar, R. Amirtharajah, P.J. Hurst, A full-wave rectifier for interfacing with multi-phase piezoelectric energy harvesters, in *IEEE International Solid-State Circuits Conference, 2008. ISSCC 2008. Digest of Technical Papers*, Feb 2008, pp. 302–615

23. G. Bawa, M. Ghovanloo, Active high power conversion efficiency rectifier with built-in dual-mode back telemetry in standard CMOS technology. IEEE Trans. Biomed. Circuits Syst. **2**(3), 184–192 (2008)

24. S, Guo, H. Lee, An efficiency-enhanced CMOS rectifier with unbalanced-biased comparators for transcutaneous-powered high-current implants. IEEE J. Solid-State Circuits **44**(6), 1796–1804 (2009)

25. F. Mounaim, M. Sawan, Integrated high-voltage inductive power and data-recovery front end dedicated to implantable devices. IEEE Trans. Biomed. Circuits Syst. **5**(3), 283–291 (2011)

26. H.-M. Lee, M. Ghovanloo, An integrated power-efficient active rectifier with offset-controlled high speed comparators for inductively powered applications. IEEE Trans. Circuits Syst. I: Regul. Pap. **58**(8), 1749–1760 (2011)

27. T.T. Le, J. Han, A. von Jouanne, K. Mayaram, T.S. Fiez, Piezoelectric micro-power generation interface circuits. IEEE J. Solid-State Circuits **41**(6), 1411–1420 (2006)

28. H.-M. Lee, M. Ghovanloo, A high frequency active voltage doubler in standard CMOS using offset-controlled comparators for inductive power transmission. IEEE Trans. Biomed. Circuits Syst. **7**(3), 213–224 (2013)

29. J.D. Chatelain, R. Dessoulavy, *Electronique - Polytechniques et universitaires romandes*, 2nd edn., 3.7.2 (Presses polytechniques et universitaires romandes, Lausanne, 1995)

30. R. Dominguez Castro, A. Rodriguez Vazquez, F. Medeiro, J.L. Huertas, High resolution CMOS current comparators, in *Eighteenth European Solid-State Circuits Conference, 1992. ESSCIRC '92*, Sept 1992, pp. 242–245

31. P. Favrat, Electronique integre pour microactionneurs electrostatiques. Ph.D. thesis, Ecole Polytechnique Federale de Lausanne, Lausanne, 1998

32. J. Shin, I.-Y. Chung, Y.J. Park, H.S. Min, A new charge pump without degradation in threshold voltage due to body effect [memory applications]. IEEE J. Solid-State Circuits **35**(8), 1227–1230 (2000)
33. J. Brown, Transmission lines at audio frequencies, and a bit of history (2008)
34. Infinity Power Solutions. Standard Product Selection Guide
35. P. Li, R. Bashirullah, A wireless power interface for rechargeable battery operated medical implants. IEEE Trans. Circuits Syst. II: Express Briefs **54**(10), 912–916 (2007)
36. P. Wang, B. Liang, X. Ye, W.H. Ko, P. Cong, A simple novel wireless integrated power management unit (PMU) for rechargeable battery-operated implantable biomedical telemetry systems, in *2010 4th International Conference on Bioinformatics and Biomedical Engineering (iCBBE)*, June 2010, pp. 1–4
37. S.-Y. Lee, M.Y. Su, M.-C. Liang, Y.-Y. Chen, C.-H. Hsieh, C.-M. Yang, H.-Y. Lai, J.-W. Lin, Q. Fang, A programmable implantable microstimulator SoC with wireless telemetry: application in closed-loop endocardial stimulation for cardiac pacemaker. IEEE Trans. Biomed. Circuits Syst. **5**(6), 511–522 (2011)
38. S.-Y. Lee, C.-H. Hsieh, C.-M. Yang, Wireless front-end with power management for an implantable cardiac microstimulator. IEEE Trans. Biomed. Circuits Syst. **6**(1), 28–38 (2012)
39. M.S.J. Steyaert, W.M.C. Sansen, Power supply rejection ratio in operational transconductance amplifiers. IEEE Trans. Circuits Syst. **37**(9), 1077–1084 (1990)
40. Texas Instrument. LDO PSRR Measurement Simplified: SLAA414 (2009)
41. J. Jung, K. Ha, J. Lee, Y. Kim, D. Kim, Wireless body area network in a ubiquitous healthcare system for physiological signal monitoring and health consulting. Int. J. Signal Process. Image Process. Pattern Recognit. **1**, 47–54 (2008)
42. C. Otto, A. Milenković, C. Sanders, E. Jovanov, System architecture of a wireless body area sensor network for ubiquitous health monitoring. J. Mob. Multimed. **1**(4), 307–326 (2005)
43. H. Li, W. Li, A high-performance ASK demodulator for wireless recovery system, in *International Conference on Wireless Communications, Networking and Mobile Computing, 2007. WiCom 2007*, Sept 2007, pp. 1204–1207
44. C.-S.A. Gong, M.-T. Shiue, K.-W. Yao, T.-Y. Chen, Y. Chang, C.-H. Su, A truly low-cost high-efficiency ASK demodulator based on self-sampling scheme for bioimplantable applications. IEEE Trans. Circuits Syst. I: Regul. Pap. **55**(6), 1464–1477 (2008)
45. H. Yu, Y. Li, L. Jiang, Z. Ji, A 31 μW ASK clock and data recovery circuit for wireless mplantable systems, in *2010 International Symposium on Intelligent Signal Processing and Communication Systems (ISPACS)*, Dec 2010, pp. 1–4
46. A. Denisov, E. Yeatman, Ultrasonic vs. inductive power delivery for miniature biomedical implants, in *2010 International Conference on Body Sensor Networks (BSN)*, June 2010, pp. 84–89
47. K. Yadav, I. Kymissis, P.R. Kinget, A 4.4 μw wake-up receiver using ultrasound data communications, in *2011 Symposium on VLSI Circuits (VLSIC)*, June 2011, pp. 212–213
48. G. Kim, Y. Lee, S. Bang, I. Lee, Y. Kim, D. Sylvester, D. Blaauw, A 695 pW standby power optical wake-up receiver for wireless sensor nodes, in *2012 IEEE Custom Integrated Circuits Conference (CICC)*, Sept 2012, pp. 1–4
49. S. Galal, B. Razavi, 10-*Gb/s* limiting amplifier and laser/modulator driver in 0.18-μm CMOS technology. IEEE J. Solid-State Circuits **38**(12), 2138–2146 (2003)
50. H.-Y. Huang, J.-C. Chien, L.-H. Lu, A 10-Gb/s inductorless CMOS limiting amplifier with third-order interleaving active feedback. IEEE J. Solid-State Circuits **42**(5), 1111–1120 (2007)
51. E.A. Vittoz, The design of high-performance analog circuits on digital CMOS chips. IEEE J. Solid-State Circuits **20**(3), 657–665 (1985)
52. J.J.F. Rijns, CMOS low-distortion high-frequency variable-gain amplifier. IEEE J. Solid-State Circuits **31**(7), 1029–1034 (1996)
53. B. Calvo, S. Celma, P.A. Martinez, M.T. Sanz, 1.8 V-100 MHz CMOS programmable gain amplifier, in *2006 IEEE International Symposium on Circuits and Systems, 2006. ISCAS 2006. Proceedings*, May 2006, 4 pp.

54. J. Chang, A.A. Abidi, C.R. Viswanathan, Flicker noise in CMOS transistors from subthreshold to strong inversion at various temperatures. IEEE Trans. Electron Devices **41**(11), 1965–1971 (1994)
55. K. Penttila, M. Keskilammi, L. Sydanheimo, M. Kivikoski, Radar cross-section analysis for passive RFID systems. IEE Proc. Microwaves Antennas Propag. **153**(1), 103–109 (2006)
56. C.-C. Yen, A.E. Gutierrez, D. Veeramani, D. van der Weide, Radar cross-section analysis of backscattering RFID tags. IEEE Antennas Wirel. Propag. Lett. **6**, 279–281 (2007)
57. S.-n. Suzuki, S.Kimura, T. Katane, H. Saotome, O. Saito, K. Kobayashi, Power and interactive information transmission to implanted medical device using ultrasonic. Jpn. J. Appl. Phys. **41**(Part 1, No. 5B), 3600–3603 (2002)
58. G.J. Saulnier, H.A. Scarton, A.J. Gavens, D.A. Shoudy, T.L. Murphy, M. Wetzel, S. Bard, S. Roa Prada, P. Das, P1G-4 through-wall communication of low-rate digital data using ultrasound, in *IEEE Ultrasonics Symposium, 2006*, Oct 2006, pp. 1385–1389
59. D.A. Shoudy, G.J. Saulnier, H.A. Scarton, P.K. Das, S. Roa Prada, J.D. Ashdown, A.J. Gavens, P3F-5 an ultrasonic through-wall communication system with power harvesting, in *IEEE Ultrasonics Symposium, 2007*, Oct 2007, pp. 1848–1853
60. Y. Hu, M. Sawan, A fully integrated low-power BPSK demodulator for implantable medical devices. IEEE Trans. Circuits Syst. I: Regul. Pap. **52**(12), 2552–2562 (2005)
61. C.-K. Liang, J.-J.J. Chen, C.-L. Chung, C.-L. Cheng, C.-C. Wang, An implantable bi-directional wireless transmission system for transcutaneous biological signal recording. Physiol. Meas. **26**(1), 83 (2005)
62. S. Mandal, R. Sarpeshkar, Power-efficient impedance-modulation wireless data links for biomedical implants. IEEE Trans. Biomed. Circuits Syst. **2**(4), 301–315 (2008)

Chapter 6
Wireless Power Transfer (WPT) and Communication

Keywords Wireless power transfer · Wireless communication · Ultrasound · CMOS · Battery · Attenuation · Transponder · Backscattering · Rectifier · Efficiency · Sensor node · Power amplifier · Uplink · Downlink · Acoustic · Human body · Health · Charge

The wireless body area network (WBAN) includes up to ten sensor nodes and an external control unit to synchronize the wireless power transfer and communication. In Chap. 4, the control unit (CU) architecture to drive a spherical array transducer is presented along with details about the front-end and back-end electronics. In Chap. 5, the transponder (TR) architecture is shown with important considerations on energy harvesting and speed limits due to transducer bandwidth. Here, the interaction between the control unit and the transponder is presented. To show advantages and limitations imposed by ultrasound (US), two prototypes for the transponder are built.

An autonomous sensor node is designed using discrete components and tested on standard FR4 substrate. The board dimensions are $2.7 \times 4.9\,\mathrm{cm}^2$ which fits inside the Kinetra neurostimulator casing (http://www.medtronic.com/downloadablefiles/220822001.pdf), wherein a six elements array transducer ($5 \times 10\,\mathrm{mm}^2$) is glued. The transponder power consumption is $794\,\mu\mathrm{A}$ when transmitting data towards the control unit, and only $12\,\mu\mathrm{A}$ when in sleep mode. A temperature sensor is available but not included in the experimental analysis.

The second sensor node merges full-custom solutions in standard CMOS process along with available discrete components. The energy harvesting interface is designed using high-voltage (HV) CMOS modules, while the receiver front-end circuit is implemented using low-voltage (LV) CMOS modules. Hence, the usage of the active area is optimized as the minimum transistor channel length can be set. To avoid damages to the LV modules, an off-chip single pole double throw (SPDT) switch is needed. For this version, one element of the array transducer is used ($5 \times 1\,\mathrm{mm}^2$); the transmitter and the sensor are not included in the design.

The chapter discusses sensor node architectures which will be called *Sensor Node I or (SN_I)* and *Sensor Node II (or SN_II)* for an easy distinction. The ultrasound

© Springer Nature Switzerland AG 2020
F. Mazzilli, C. Dehollain, *Ultrasound Energy and Data Transfer for Medical Implants*,
Analog Circuits and Signal Processing, https://doi.org/10.1007/978-3-030-49004-1_6

link efficiency and the reliability of data communication are measured. Lastly, a
summary of the chapter is given.

6.1 Sensor Node I

Figure 6.1 shows the block diagram of the Sensor Node I realized using discrete
components. The voltage swing of the piezoelectric transducers is boosted thanks to
a parallel inductor and to avoid over voltage is limited to ± 10 V through two zener
diodes. The Atmel ATMEGA48A microcontroller is the core of the sensor node
which manages the different tasks as energy harvesting, communication in uplink
and downlink, and the sensor activity. The load shift keying (LSK) modulator,
which was presented in Sect. 5.7.2, is always connected to the array transducer,
but is enabled via the microcontroller through signals En_Comm_PMOS and
En_Comm_NMOS. To allow the recharge of the microbattery the reference
voltage V_{ref} for the piezo transducers is by default set to ground (see switch 1 in
Fig. 6.2). The switch SW_1 is also ON during the charging phase and the half-wave
rectifier can generate a DC voltage equal to V_{rec} which is subsequently regulated to
V_{reg} using the National Semiconductor LP2980-ADJ low dropout (LDO) adjustable
voltage regulator. The Infinite Power Solutions (IPS) microbattery requires a
tight constant voltage to be recharged [1], thus the Linear Technology LTC1540
nanopower comparator ($Comp_1$) checks that V_{reg} is equal to 4.1 V before turning
on the switch SW_2 so that current can flow from the transducer to the microbattery.
The Micrel MIC841A micropower voltage comparator ($Comp_2$) checks that the
microbattery voltage V_{BATT} stays between V_H (4.1 V) and V_L (2.5 V). If it happens
that V_{BATT} falls out of this range, the microbattery is disconnected from the
microcontroller. The details about SW_2–SW_3 are given in Sect. 5.2. The supply
voltage for the microcontroller is set to 3.3 V using the MICREL MIC5231 low
dropout voltage regulator. While charging, the microcontroller can be turned on and

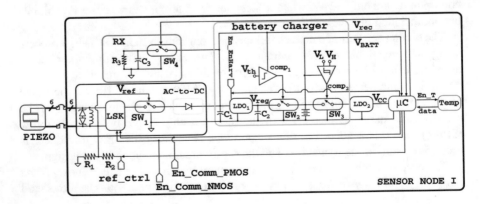

Fig. 6.1 Block diagram of the Sensor Node I

Fig. 6.2 Circuit representations of the switches

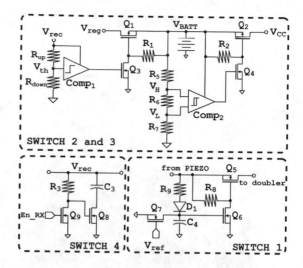

Table 6.1 Values of the passive components

Component	Value
R_1, R_2	2 (MΩ)
R_3, R_9, R_{down}	100 (kΩ)
R_5, R_8	1 (MΩ)
R_6	387 (kΩ)
R_7	606 (kΩ)
R_{up}	270 (kΩ)
C_1	10 (nF)
C_2, C_3, C_4	2.2 (μF)
L	100 (μH)

check the rectifier voltage level, the battery status and the output of the MAXIM MAX6576 temperature sensor which consumes 140 μA.

Figure 6.2 shows the implementation of the switches SW_1–SW_4 with MOSFETs. The switch SW_1 is normally ON since it is self-biased through the limiting resistor R_9 and the passive rectifier composed by D_1 and C_4. To stop the harvesting phase, V_{ref} increases from 0 V up to $V_{CC}/2$, thus Q_5–Q_6 are turned off. The signal En_RX is set to ground by the microcontroller, so Q_8 conducts and C_3 is in parallel with C_1. In case the signal En_RX is set to V_{CC}, R_3 is in parallel with C_1. By a proper sizing of the passive components (e.g. $C_3 \gg C_1$), the charging of the battery or the envelope detection of the incoming signal can be performed.

Table 6.1 summarizes values of passive components used in the sensor nodes. The battery needs to be continuously monitored to prevent any chemical damage, thus an average current of 5 μA is sinked by $Comp_2$ and the equivalent resistance $R_5 + R_6 + R_7 \approx 2\,M\Omega$.

The Sensor Node I is powered by ultrasound and once the battery is charged, the system needs to be addressed before transmitting the collected sensor data. To

Fig. 6.3 State chart of the Sensor Node I

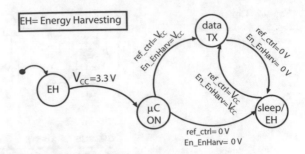

PCB 4 layers
size: 2.7 cm x 4.9 cm
substrate: FR4

Fig. 6.4 Discrete implementation of the implanted medical device

simplify the software for the microcontroller, the data transmission starts as soon as the battery is charged, thus the addressing phase is omitted. To test load modulation via acoustic waves, the sensor data are replaced by a pseudorandom sequence of 32 signal elements. To synchronize the data pattern between the transmitter and the receiver in the control unit, 15 signal elements are added to the pseudorandom sequence.

Figure 6.3 shows the state chart of the Sensor Node I when the supply voltage V_{cc} for the microcontroller reaches 3.3 V. The sensor node is either transmitting data or in sleep mode according to the signals ref_ctrl and En_EnHarv controlled by the microcontroller, which default values are set to 0 V to allow energy harvesting.

Figure 6.4 shows the Sensor Node I, placed inside a biocompatible casing manufactured by MEDTRONIC (http://www.medtronic.com/downloadablefiles/

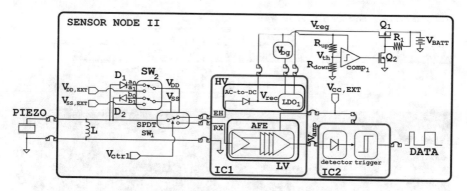

Fig. 6.5 Block diagram of the Sensor Node II

220822001.pdf) for in-vivo test. The array transducer is fabricated by IMASONIC and is provided with six signal pins and two reference pins.

However, the sensor node is tested with the transducer shown in Fig. 3.14 (Chap. 3) since only in-vitro test is performed.

To study the possibility to transmit data toward an implanted medical device from an external control unit, a second sensor node is designed which includes the main energy harvesting block as the AC-to-DC converter and the LDO to recharge the microbattery.

6.2 Sensor Node II

Figure 6.5 shows the block diagram of the Sensor Node II realized using discrete components and two full-custom integrated circuits (ICs) in a standard CMOS process. The integrated circuit labeled as *IC1* is presented in this book, and the results were presented in Chap. 5. The integrated circuit labeled as *IC2* is reused from another work and it is required to validate the reliability of the up-link communication. The architecture of *IC2* is presented in Sect. 6.2.1. One element of the flat array transducer is connected to the sensor node, a parallel inductor L is used to boost the output voltage swing of the piezo. As the *IC1* integrates the analog-front-end (AFE) receiver (RX) using low-voltage (LV) CMOS modules and the energy harvesting (EH) blocks (AC-to-DC converter and LDO) using high-voltage (HV) CMOS modules, the Analog Device ADG1419 SPDT switch (SW_1) is used to set either the charge of the microbattery or the uplink communication. The SPDT is self-biased via diodes D_1–D_2 during the EH phase, and it is externally biased via $V_{DD,EXT}$ and $V_{SS,EXT}$ when in the RX phase. The required positive and negative supply voltages are equal to $V_{DD} = 5$ V and $V_{SS} = -5$ V. The Sensor Node II does not include a microcontroller, thus any control signal is manually set. A dip switch SW_2 is needed to select the right supply voltage source during the EH and RX

Table 6.2 Logic table for SW_1 and SW_2

SW_1			SW_2					
V_{ctrl}	EH	RX	a_0	a_1	b_0	b_1	V_{DD}	V_{SS}
0	ON	OFF	1	0	1	0	Self-biased	Self-biased
1	OFF	ON	0	1	0	1	$V_{DD,EXT}$	$V_{SS,EXT}$

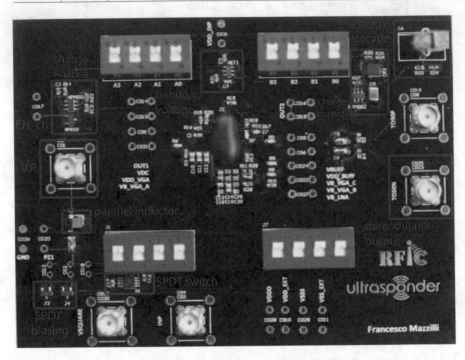

Fig. 6.6 Board to test the integrated implementation of the implanted medical device

phase for SW_1, respectively. The voltage regulator presented in Sect. 5.5 requires a reference voltage V_{bg} to provide a constant output voltage $V_{reg} = 4.1$ V. Hence, the Linear Technology LT6656 precision voltage reference provides a $V_{bg} = 1.25$ V while being powered by the AC-to-DC converter via V_{rec}. The remaining part of the sensor node was previously discussed in Fig. 6.2.

Table 6.2 represents the output status for SW_1 and SW_2 according to the control signals V_{ctrl}, a_0–a_1 and b_0–b_1.

Figure 6.6 shows the integrated implementation of the implanted medical device, where *IC1* is in the center of the board and *IC2* is not shown here. The energy storage element is represented by a supercapacitor, but can be replaced by the IPS microbattery. The ultrasound input is connected to the connector marked with V_P, the unused SMA connectors are not highlighted in dark red.

In the following section, the circuit description for the *IC2* is given.

6.2.1   IC2

Figure 6.7 shows the architecture of the demodulator integrated in the *IC2*, which is composed of four subcircuits, including an envelope detector, a digital shaper, a Schmitt Trigger, and a load driver. The proposed architecture optimizes the circuit in [2] adding the Schmitt Trigger described in [3] to overcome sudden changes at the output of the digital shaper.

Figure 6.8 shows the schematic of the demodulator integrated in the *IC2*, which consumes an average power of $6\,\mu$W at $V_{CC} = 1.5$ V. The output of the analog front end V_{amp} is fed into the half-wave rectifier m_1–m_5, whose output voltage $V_{envelope}$ is tracked by the peak detector C_1.

The digital shaper converts the envelope signal into a digital signal V_{shaper}, the minimum detectable amplitude can be expressed as [2]

$$\Delta V_{envelope} = V_H - V_L = \sqrt{\frac{\beta_{10}}{\beta_9}}\left(V_H - 2V_{t,n}\right) \tag{6.1}$$

where V_H represents the high-voltage value of $V_{envelope}$, V_L the low-voltage value, β the transistor gain factor, and V_t the transistor threshold voltage. The Schmitt Trigger removes small unwanted transitions in V_{shaper} according to the low

Fig. 6.7 Architecture of the *IC2*

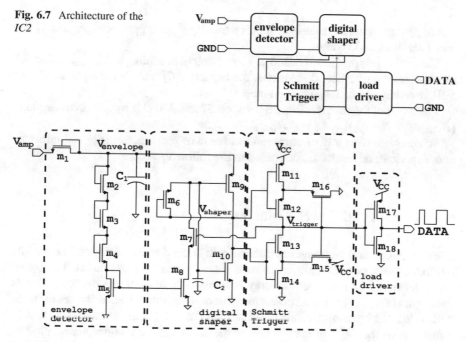

Fig. 6.8 Schematic of the *IC2*

Table 6.3 Design variables

Transistor	W (μm)	L (μm)	W/L
m_1	1.92	0.18	10.66
m_2–m_6	0.24	0.18	1.3
m_7–m_8	0.24	6	0.04
m_9	0.24	1.1	0.22
m_{10}	0.24	8	0.03
m_{11}–m_{12}	0.24	0.18	1.3
m_{13}–m_{14}	1.44	0.18	8
m_{15}	1.44	0.18	8
m_{16}	0.24	0.18	1.3
m_{17}	0.5	0.18	2.8
m_{18}	0.24	0.18	1.3

$V_{L,shaper}$ and high $V_{H,shaper}$ threshold. Thus, the transistor can be sized as follows: [3]

$$\frac{\beta_{14}}{\beta_{15}} = \left(\frac{V_{CC} - V_{H,shaper}}{V_{H,shaper} - V_{t,n}}\right)^2 \tag{6.2a}$$

$$\frac{\beta_{11}}{\beta_{16}} = \left(\frac{V_{L,shaper}}{V_{CC} - V_{L,shaper} - |V_{t,p}|}\right)^2 \tag{6.2b}$$

where β_i represents the transistor gain factor, V_{CC} the supply voltage, and V_t the transistor threshold voltage.

The load driver provides a rail-to-rail logic output according to $V_{trigger}$. Table 6.3 summarizes the transistor dimensions, the capacitor C_1 and C_2 are equal to 15 and 5 pF, thus they can be integrated on-chip.

The speed of the circuit is limited between 57 and 440 kHz by two time constants $\left(R_{ON,m_6} + R_{ON,m_9}\right) \times C_2$ and $R_{ON,m_1} \times C_1$.

To complete the study of ultrasound as mean to propagate energy and information, experimental results are discussed in the following section.

6.3   Ultrasound Power Transfer

The external control unit drives a spherical array transducer made up of 64 elements through class-E power amplifiers to send energy towards sensor nodes. The sound beam generated by the spherical array is focused at 10.5 cm and the focal area is 2×5 mm^2 (SA_{focal}) when all power amplifiers are enabled. Hence, the sensor node will be located by default at this position thus the link efficiency can be studied as a function of the DC power delivered at the output of the rectifier versus the DC power consumed by power amplifiers that are enabled. The implanted transducer has a total active area of $5 \times 10\,mm^2$ (SA_{tot}) and the active area per element of the array is $5 \times 1\,mm^2$ (SA_{el}). Since SA_{focal} is almost equal to SA_{el}, one element of the flat array transducer is connected to the rectifier. The discrete and the integrated rectifier, previously introduced, are used to compare the results.

6.3.1 Link Efficiency

Figure 6.9 shows the block diagram of the energy harvesting chain used to study the link efficiency as a function of the number of enabled power amplifier and delivered power to the load of the rectifier.

To avoid excessive power consumption for the power amplifier, the supply voltage V_{CC} is set to the minimum which is 2 V, as a result the drain efficiency is 66% (see Chap. 4). For simplicity of the analysis, the ultrasound box includes the transformation from electric to ultrasonic energy and back from ultrasonic to electric. A discrete half-wave rectifier (a diode and a capacitor) and the active rectifier, presented in Chap. 5, are used for this study.

Figure 6.10 shows the block diagram of the integrated (IC) rectifier and the circuit diagram of the half-wave rectifier. The load capacitor C_1 and C_2 are equal to 100 nF. The discrete rectifier employs the Fairchild MMBD4148SE diode with a forward voltage of 700 mV for a forward current of 10 mA. While the integrated rectifier is designed in a high-voltage 0.18 CMOS process, threshold voltages are -1.42 and 1.55 V for a PMOS and a NMOS transistor.

According to Fig. 6.9, the link efficiency can be expressed as

$$\eta_{link} = \eta_{PA} \times \eta_{US} \times \eta_{rec} \tag{6.3}$$

where η_{PA} represents the drain efficiency (DE) of the power amplifier, η_{US} the conversion from electric to ultrasound and from ultrasound to electric, and η_{rec} the rectifier efficiency. Thus, the link efficiency is strongly dependent from η_{US} as in the case of inductive link. However, (6.3) does not consider the effect of misalignment between the spherical and flat transducer as well as the shape of the ultrasound beam.

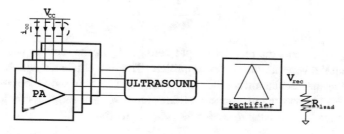

Fig. 6.9 Block diagram of the energy harvesting chain

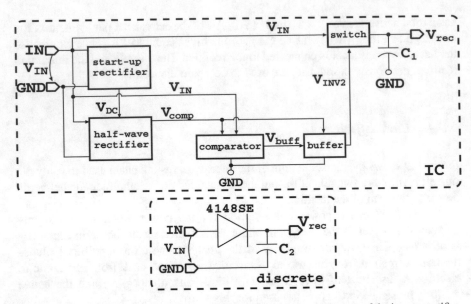

Fig. 6.10 Block diagram of the integrated (IC) rectifier and circuit diagram of the iscrete rectifier

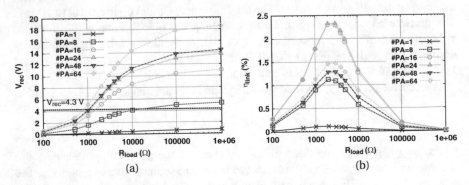

Fig. 6.11 Measured: (**a**) rectifier output voltage and (**b**) link efficiency versus R_{load}

To simplify the analysis, the link efficiency can be also written as

$$\eta_{link} = \frac{V_{rec}^2}{R_{load}} \times \frac{1}{V_{cc}i_{cc}} \tag{6.4}$$

where V_{rec} represents the rectifier output voltage, R_{load} the rectifier load, i_{CC} the total current delivered to the power amplifiers. The link efficiency is first analyzed in water, thus the attenuation coefficient 0.0231 dB at 10.5 cm should be considered.

Figure 6.11 shows the measured rectifier output voltage and link efficiency as a function of the load resistor R_{load} for the discrete rectifier.

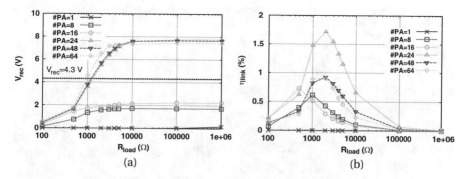

Fig. 6.12 Measured: (**a**) rectifier output voltage and (**b**) link efficiency versus R_{load}

Table 6.4 Table of acoustic attenuation α for human tissue at 1 MHz

Material	α (dB/cm)
Breast	0.75
Fat	0.48
Muscle	1.09
Soft tissue (average)	0.54
Water	0.0022

Water is listed for reference

Figure 6.12 shows the measured rectifier output voltage and link efficiency as a function of the load resistor R_{load} for the integrated rectifier.

The link efficiency is higher when the discrete rectifier is used in the energy harvesting chain as the threshold voltage of the diode is lower than the threshold voltage of the CMOS transistor. The integrated circuit employs a self-biased comparator which may affect the rectifier output voltage thus decreasing the link efficiency.

The link efficiency is plotted as a function of the number of power amplifiers that are enabled for the transmission of power. A maximum link efficiency of 2.3% and 1.54% is measured for the discrete and the IC rectifier, while 24 power amplifiers are turned on and the R_{load} is equal to 3000 kΩ, roughly equal to the transducer resistance $R_P = 3510$ kΩ. An inductor of 100 μH is also used to boost the rectifier input voltage and to compensate the capacitor of the transducer.

Since attenuation coefficients for human tissues are higher than ultrasound attenuation coefficient in water [4], a phantom material is developed by INSERM to mimic the human tissue [5]. The measured attenuation of the phantom material is $\alpha_P = 0.33$ dB cm^{-1} MHz^{-1}. Table 6.4 lists the acoustic attenuation for human tissue at 1 MHz, water is given for reference.

Figure 6.13 shows the in-vitro set-up used to test the wireless power transfer: the spherical array transducer is on the left, the flat array transducer on the right, and the phantom material in between. The distance between the two transducers is 10.5 cm, the length of the phantom material is 9.5 cm, and the remaining 1 cm is water.

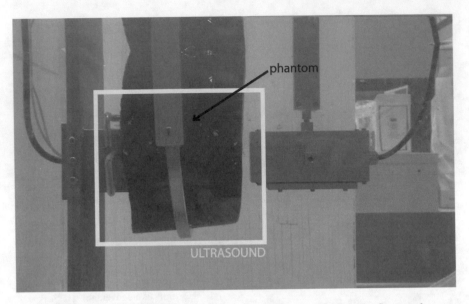

Fig. 6.13 In-vitro test set-up showing the spherical transducer on the left, the flat transducer on the right, and the phantom material in the middle

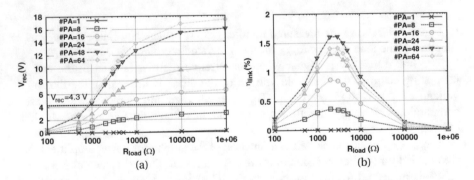

Fig. 6.14 Measured: (**a**) rectifier output voltage and (**b**) link efficiency versus R_{load}

Figure 6.14a shows the measured rectifier output voltage and Fig. 6.14b the link efficiency as a function of the load resistor R_{load} for the discrete rectifier.

Figure 6.15a shows the measured rectifier output voltage and Fig. 6.15b the link efficiency as a function of the load resistor R_{load} for the integrated rectifier.

A maximum link efficiency of 1.60% and 1.03% is measured for the discrete and the IC rectifier for R_{load} in between 2000 and 3000 kΩ. In addition, the maximum is found when 48 power amplifiers are enabled.

To compare the results with and without the phantom material, the ultrasound beam must be analyzed. However, when the phantom material is placed between the transducers the maximum link efficiency decreases roughly by one third.

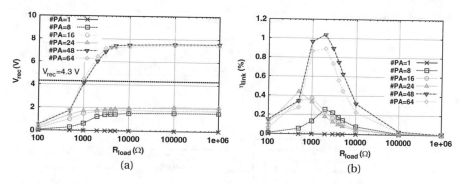

Fig. 6.15 Measured: (**a**) rectifier output voltage and (**b**) link efficiency versus R_{load}

6.4 Ultrasound Communication

In this section, the ultrasound wireless communication is discussed.

The external control unit controls the transfer of energy as well as the wireless communication in uplink (return link), and downlink (forward link). During the downlink, the control unit can address a sensor node by duty cycling the input signal to the gate of the power MOSFET within a power amplifier. While during the uplink, the control unit uses m power amplifiers to transmit a continuous wave towards the sensor node, which modulates the load of its transducer to send back information (e.g. sensor activity). Then, the control unit selects n low-noise amplifiers to demodulate the reflected wave generated by the sensor node.

Three different control units to sensor node operation modes can be defined: (1) energy harvesting, (2) uplink and (3) downlink. During the energy harvesting mode, the control unit sends enough energy towards the sensor node to recharge the microbattery in maximum 30 min. The recharge status is checked by the microcontroller, which enables the load modulator and sends the information through the uplink to the control unit. Hence, the control unit can adjust the supply voltage to the power amplifiers and address the wanted sensor node through the downlink. Lastly, the uplink mode is used to retrieve the sensor information. The control unit adjusts the supply voltage in order to keep the bit error rate (BER) below a certain threshold (e.g. 10^{-3}).

Table 6.5 summarizes the operation modes between the control unit and the sensor node, and the evolution of the power amplifier (PA) supply voltage.

In this work, the power supply and the number of active power amplifiers as well as the number of active low-noise amplifiers are manually configured to achieve the best performance.

Table 6.5 Operation modes

Mode	Action	Operation	PA supply
Mode 1	Recharge	Energy harvesting	V_{CC} ↗
Mode 2	Addressing	Downlink	V_{CC} ↘↘
Mode 3	Measurement	Uplink	V_{CC} ↘

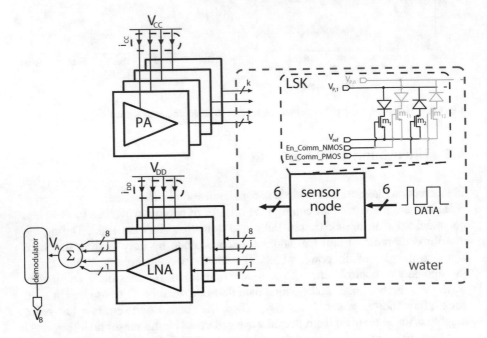

Fig. 6.16 Block diagram of the uplink

6.4.1 Uplink

Figure 6.16 shows a simplified block diagram of the uplink communication, wherein the *Sensor Node I* sends data using backscattering. The control unit sends a continuous wave at 1 MHz towards the transponder, and the microcontroller enables the load modulator (*LSK*) through signals *En_Comm_NMOS* and *En_Comm_PMOS* according to the stored data.

As shown in [6], the presence of standing waves becomes negligible when the distance between the receiver and the transmitter is equal to 10 cm. Moreover, the difference in reflected power is equal to 6% when the implanted transducer is either matched or connected to a 10 kΩ resistor. While the percentage of reflected power increases drastically when the transducer is short circuited. Hence, to simplify the modulator architecture, the piezoelectric transducer is either short circuited or left unloaded. In this manner, the ratio between the absorbed and reflected power determines the modulation index at the control unit receiver. The data rate can be chosen between 5 and 20 kbps in order to decrease the microcontroller power consumption.

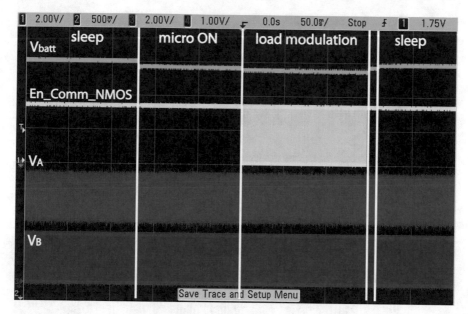

Fig. 6.17 Data transfer during the uplink

Figure 6.17 shows the uplink operation; when the microcontroller turns on the battery voltage starts to drop and as soon as the data transfer starts the battery voltage drops further. The power consumption during the sleep mode is 12 μA and the average power consumption during the data transfer is 794 μA (*micro ON* and *load modulation*).

For simplicity, the uplink communication is tested without the effect of the phantom material, and the transponder placed at 10.5 cm from the spherical transducer surface. To maintain the link efficiency to its maximum, 16 power amplifiers are enabled (see Fig. 6.11b) and the supply voltage is set to 3 V to recharge the battery in less than 30 min. Once the data transfer starts, the presence of standing waves causes the bit error rate (BER) to increase. By decreasing the supply voltage of the power amplifiers down to 2 V, the BER decreases.

Figure 6.18 shows the signal waveforms *En_Comm_NMOS* and V_B during load modulation. The data stream is represented by a synchronization part followed by the sensed data. The data rate is set to 10 kbps since Manchester encoding is implemented by the microcontroller. The synchronization signal and the data are equal to 111101 and 0011001100110011.

Figure 6.18a and b represent the signal V_B acquired with two- and three-elements of the spherical array transducer. The quality of the communication improves as the number of receiving channels increases. The measured modulation index is equal to 2.7% and is measured at V_A.

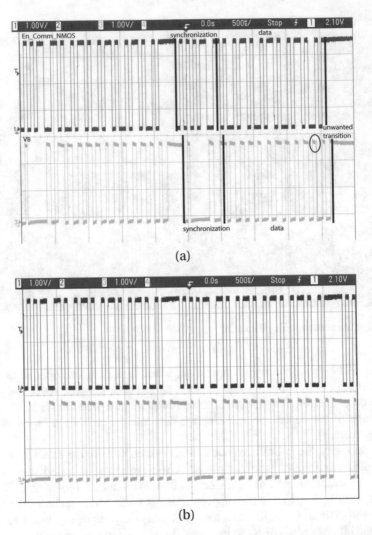

Fig. 6.18 Measured signal *En_Comm_NMOS* and V_B for: (**a**) two receiving channels and (**b**) three receiving channels

Figure 6.19 shows the output spectrum at V_A normalized to the carrier frequency (1 MHz) for different receiving channels measured with the Agilent 4394A spectrum analyzer (resolution bandwidth equal to 100 Hz). At low side bands, the presence of two tones at 993 and 989 kHz decreases the quality of the uplink communication when the difference in power is less than 5 dB.

Figure 6.19a plots the output spectrum when one and two receiving channels are used to demodulate the incoming message. As shown in Fig. 6.18a, errors may occur during the deconding process and this is due to the presence of spurious tones. As the modulator is switch based, the generation of harmonics may occur. Moreover, the

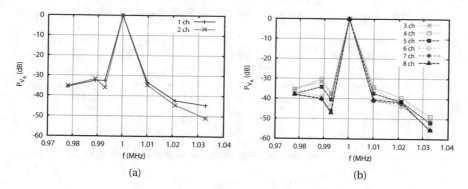

Fig. 6.19 Measured spectrum at V_A normalized to the carrier frequency for different receiving channels

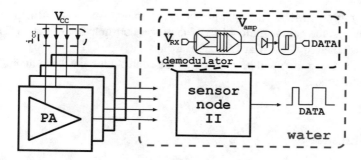

Fig. 6.20 Block diagram of the downlink

Sensor Node I employs all channels available to transmit the data, thus the strength of the unwanted harmonics can increase.

Figure 6.19b plots the output spectrum when more than two channels are used to demodulate the incoming message. In this situations, the quality of communication improves as the spurious tone at 989 kHz decreases.

The load modulation through ultrasound is demonstrated; in the next subsection the downlink communication is discussed.

6.4.2 Downlink

Figure 6.20 shows a simplified block diagram of the downlink communication, wherein the *Sensor Node II* demodulates the incoming message. The demodulator is tested with laboratory equipment since the transmitter is not implemented in the control unit yet.

Figure 6.21 shows the input and the output voltage waveforms of the demodulator for 40% and 100% modulation index and a data rate equal to 50 kbps and carrier

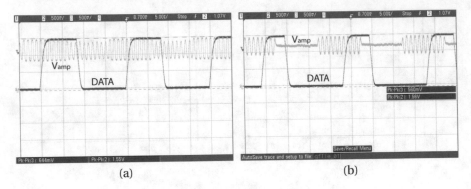

Fig. 6.21 Measured input and the output voltage waveforms of the demodulator

frequency $f_0 = 1$ MHz. The input power to the analog front end (AFE) is equal to -55 dBm and the gain is set to 45 dB. The power consumption of the AFE and the decoder is equal to 184 μW at 1.5 V.

Once the transmitter will be available in the control unit, the demodulator will be tested within the in-vitro set-up.

6.5 Summary

The design of two sensor nodes to test remote powering and wireless communication through ultrasound has been presented in this chapter.

Firstly, the energy harvesting chain has been characterized, which included m power amplifiers, the conversion from electrical to ultrasound power and from ultrasound to electrical power, and the rectifier. The number of power amplifiers has been swept to find the maximum link efficiency, while only an element of the implanted transducer has been used to harvest energy. The maximum link efficiency has been equal to 2.3% and 1.5% in water when using a discrete and an integrated rectifier. However, the link efficiency has been dropped down to 1.6% and 1.03% when a phantom material has been placed between the transmitting and the receiving transducers. Without the phantom the optimum number of power amplifiers has been in between to 24 and 16. While in presence of the phantom material the optimum number of power amplifiers has been increased to 48.

Lastly, the uplink and downlink communication have been tested. The quality of the uplink communication has been shown that can be improved by increasing the number of receiving elements. Moreover, the demodulation has been successfully demonstrated with a data rate equal to 10 kbps and a modulation index of 2.7%. In addition, the downlink working principle has been shown with the aid of laboratory equipment. The power consumption has been 184 μW at 1.5 V.

References

1. Infinity Power Solutions. Standard Product Selection Guide
2. C.-C. Wang, C.-L. Chen, R.-C. Kuo, D. Shmilovitz, Self-sampled all-MOS ASK demodulator for lower ISM band applications. IEEE Trans. Circuits Syst. II: Express Briefs **57**(4), 265–269 (2010)
3. I.M. Filanovsky, H. Baltes, CMOS Schmitt trigger design. IEEE Trans. Circuits Syst. I: Fundam. Theory Appl. **41**(1), 46–49 (1994)
4. M.O. Culjat, D. Goldenberg, P. Tewari, R.S. Singh, A review of tissue substitutes for ultrasound imaging. Ultrasound Med. Biol. **36**(6), 861–873 (2010)
5. J. Oudry, C. Bastard, V. Miette, R. Willinger, L. Sandrin, Copolymer-in-oil phantom materials for elastography. Ultrasound Med. Biol. **35**(7), 1185–1197 (2009)
6. S. Arra, J. Leskinen, J. Heikkila, J. Vanhala, Ultrasonic power and data link for wireless implantable applications, in *ISWPC '07. 2nd International Symposium on Wireless Pervasive Computing, 2007*, Feb 2007

Chapter 7
Conclusion

Passive radio-frequency identification (RFID) transponders, while available for many years, have only recently been applied to humans. These transponders are encoded and implanted in a patient, and subsequently accessed with a hand-held electromagnetic reader in a quick and non-invasive manner. However, while an external RF transmitter could be used to communicate with the transponder, RF and magnetic energy may only penetrate a few centimeters into a body, because of its dielectric nature. To be able to communicate effectively with a transponder that is located deep within the body and providing wireless power transfer to a microbattery, ultrasonic waves are used as a valid alternative.

In this book, ultrasound waves have been extensively studied as new approach for wireless energy transfer and communication dedicated to battery-powered implanted medical devices. An in-vitro set-up has been built to perform primary analysis of the link budget and to estimate the performance of data transmission. The tank has been made of Plexiglas and occupies a volume of $50 \times 50 \times 50\,cm^3$, while a mechanical system has been used to hold a pair of acoustic transducers.

Exposure limits to ultrasound for human tissues have been introduced and system specifications have been derived. An operating frequency of 1 MHz has been selected to minimize bioeffects (e.g. cavitations) and to maximize the width of the ultrasound beam and the acoustic intensity at the implant location. A thin-film Lithium Phosphorus Oxynitride (LiPON) rechargeable battery from Infinite Power Solutions has been chosen to provide power to the implanted device. A spherical array transducer made up of 64 elements with a total active area of $30 \times 96\,mm^2$ has been selected to build the ultrasound link. The array transducer overcomes issue as misalignment and can deliver higher power compared to single element transducers. Moreover, a flat array transducer made up of six elements with a total active area of $5 \times 10\,mm^2$ has been used for the implanted device. Lastly, the choice of the CMOS process to design the implant front-end electronics has been pursed. The

© Springer Nature Switzerland AG 2020

F. Mazzilli, C. Dehollain, *Ultrasound Energy and Data Transfer for Medical Implants*,
Analog Circuits and Signal Processing, https://doi.org/10.1007/978-3-030-49004-1_7

XFAB 0.18 μm high-voltage CMOS process has been selected since offers high power modules with higher oxide breakdown voltage.

To enable the operations with the implant as energy transfer and data communication, an external control has been realized in discrete components. Two field programmable gate arrays (FPGAs) have been used as central processing unit to synchronize the operations. To send energy and data, a class-E power amplifier has been designed to drive the spherical array. A maximum drain efficiency (DE) and power added efficiency (PAE) of 70% and 57% have been measured when the supply voltage is 3 V. An acoustic intensity of 23 W/cm^2 at 10.5 cm has been recorded in water when the ultrasound transducer has been connected to the power amplifiers. The electrical power consumption due to the amplifier was equal to 8 W. To retrieve the sense data by the implant, a receiver has been designed which employs seven amplifiers to be able to detect voltage amplitudes in the order of few milliVolts.

To increase the power conversion efficiency from ultrasound to electrical energy and to decrease the power consumption of the implant when in active mode, considerable work has been done towards these directions. First, an equivalent model of the ultrasound piezoelectric transducer has been introduced and validated through simulated and measured results. Next, a novel synchronous rectifier and a two-stage low-drop-out regulator, main building blocks for the recharge of the microbattery, have been designed and fabricated in standard 0.18 μm HV CMOS technology. A maximum power conversion efficiency of 82.45% has been measured for the rectifier while delivering a current of 5 mA. The regulator has been added to provide 4.1 V constant voltage to recharge the microbattery, and to decrease the noise level from the rectifier output voltage. To limit the number of recharging cycles and to preserve the autonomy of the rechargeable microbattery, a low-power receiver architecture has been envisaged which can be activated only when necessary by the implant. The front-end amplifier has been successfully characterized while consuming 178 μW at 1.5 V.

Lastly, two prototypes of the implant to demonstrate energy transfer and data communication have been proposed. A discrete version has been used to test the recharge of the microbattery via ultrasound, and the uplink communication via load modulation. The second version integrates the CMOS architectures along with discrete components to optimize the power conversion efficiency and the power consumption.

7.1 Outlook

The proposed work opens new perspectives for future research. First of all, the implant system can be fully integrated in standard CMOS process decreasing the number of external components. The output voltage of the variable gain amplifier can be implemented on-chip using an automated control loop. In addition, the use of the battery can be avoided by powering the system via a second mechanical harvester which converts for example, heartbeats into power. Lastly, to increase the angular acceptance of the control unit, a flat array transducer structure can be envisaged using a lower number of elements. In this manner, standard beamforming and beam steering techniques for phased linear array can be used.

Appendix

```
width = 1.4e-3;
height = 5e-3;
kerf = 0.1e-3;
f0=1e6;
elementsx=64;
elementsy=1;
xdc = createRectPlanarArray(elementsx,elementsy,width,height,kerf,0,[0 0 0]);
defineMedia;
lambda = water.soundspeed/f0;
xmin = -(elementsx/2 + 1) * (width + kerf);
xmax = (elementsx/2 + 1) * (width + kerf);
ymin = 0;
ymax = 0;
zmin = 0;
zmax = elementsx * 2 * f0/1e6 * lambda;
focusx=0;
focusy=0;
focusz=0.1;
xpoints = 246*4;
ypoints = 1;
zpoints = 627;
dx = (xmax-xmin)/xpoints;
dy = (ymax-ymin)/ypoints;
dz = (zmax-zmin)/zpoints;
x = xmin:dx:xmax;
y = ymin:dy:ymax;
z = zmin:dz:zmax;
delta = [dx dy dz];
cg = setCoordinateGrid(delta, xmin, xmax, ymin, ymax, zmin, zmax);
```

© Springer Nature Switzerland AG 2020
F. Mazzilli, C. Dehollain, *Ultrasound Energy and Data Transfer for Medical Implants*,
Analog Circuits and Signal Processing, https://doi.org/10.1007/978-3-030-49004-1

This sets up our coordinate grid to cover the full width of the transducer array in the x direction and to measure the pressure field to 50 wavelengths in the z direction. Now we need to focus the transducer array.

```
xdc = findSingleFocusPhase(xdc, focusx, focusy, focusz, water, f0, 200);
ndiv = 15;
pCw=fnmCall(xdc, cg, water, ndiv, f0, 0);
figure();
maxPCW=max(abs(pCw(493,1,:)));
h = pcolor(z*100,x*100,squeeze(abs(pCw))/max(maxPCW));
set(h,'edgecolor','none');
hc = colorbar;
set(hc,'FontSize',15)
title('Pressure Field at y = 0 cm, f0=1 MHz');
xlabel('z (cm)');
ylabel('x (cm)');
maxPCW=max(abs(pCw:,1,328)));
figure()
plot(x(493:527)*1000,squeeze(abs(pCw(493:527,1,328)))/max(maxPCW))
```

Index

A

Acoustics
    attenuation, 131
    intensity, 10, 52, 53, 141
    pressure, 29, 32, 63, 64
    transducers, 141
    waves, 124
Active rectifier
    ABB technique, 80–81
    available peak current, 83, 84
    available peak voltage, 84
    benchmarking, 87, 88
    bias-flip rectifier, 76–77
    biasing currents, 80
    CMOS voltage doublers, 76
    DC bias point, 79–80
    design variables, 80
    high-voltage transistors, 78, 80
    HV CMOS process, 89
    LIPON battery, 89
    lumped parameters, 81
    multi-stage/VM, 76
    on-resistance, 79
    PCE, 82–87
    phase shifting, 82
    PMOS transistors, 87
    power and output voltage, 87
    simulation and measurement, 84–86
    single-ended synchronous, 77
    single-stage active rectifier, 77–78, 81, 82
    sketches, 83
    transistors size, 80
    VCE, 82, 84–85, 87
    voltage doubler, 76
    voltage ripple, 79
    wireless micro-sensor networks, 76
Adaptive-body-bias (ABB) technique, 80–81
ADC, *see* Analog-to-digital converter (ADC)
Advanced Design System (ADS), 49
AM, *see* Amplitude modulation (AM)
Amplitude modulation (AM), 9
Amplitude-shift keying (ASK)
    low-power, 95
    and OOK demodulator, 95–109
Analog demodulation, 55, 56
Analog-front-end (AFE), 125, 138
Analog-to-digital converter (ADC), 1, 2, 45
Array transducer, 141
ASK, *see* Amplitude-shift keying (ASK)
Atmel ATMEGA48A microcontroller, 122
Attenuation, 1, 14, 20, 64, 91
Autonomous sensor node, 121

B

Backscattering, 9, 11, 109, 110
Bandwidth (BW), 63, 97, 103–106, 108, 116
Battery, 2, 8, 12, 23, 25
    block diagram, 69
    charger, 61
    high-/low-voltage battery detection circuits, 115
    IPS, 62, 89
    leakage, 90
    LIPON, 89
    microbattery, 68–69, 115
    micro-cell, 92
    recharge, 61, 95
    RFID systems, 76
Battery-powered implanted medical devices., 141

© Springer Nature Switzerland AG 2020
F. Mazzilli, C. Dehollain, *Ultrasound Energy and Data Transfer for Medical Implants*,
Analog Circuits and Signal Processing, https://doi.org/10.1007/978-3-030-49004-1

Printed in the United States
by Baker & Taylor Publisher Services